Möbius and his band

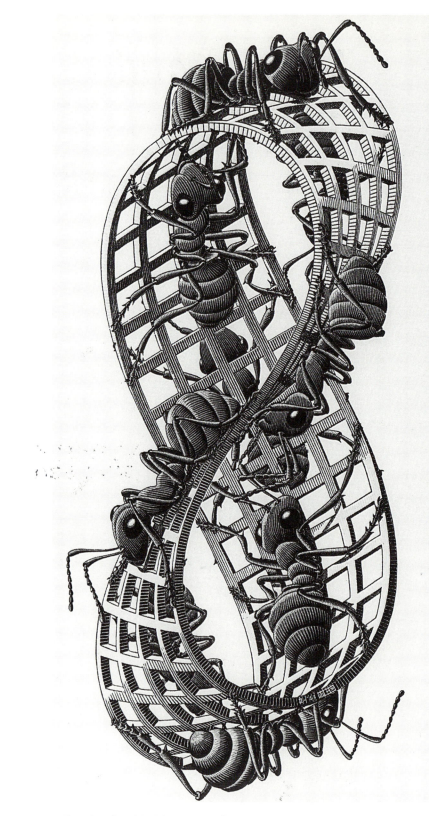

Möbius band II, 1963 by M. C. Escher.

Möbius and his band

Mathematics and Astronomy in Nineteenth-century Germany

EDITED BY

JOHN FAUVEL, RAYMOND FLOOD, and ROBIN WILSON

Oxford   New York   Tokyo
OXFORD UNIVERSITY PRESS
1993

Oxford University Press, Walton Street, Oxford OX2 6DP

Oxford   New York   Toronto
Delhi   Bombay   Calcutta   Madras   Karachi
Kuala Lumpur   Singapore   Hong Kong   Tokyo
Nairobi   Dar es Salaam   Cape Town
Melbourne   Auckland   Madrid
and associated companies in
Berlin   Ibadan

Oxford is a trade mark of Oxford University Press

Published in the United States
by Oxford University Press Inc., New York

A catalogue record for this book is available from the British Library.

Library of Congress Cataloging in Publication Data

(Data available on request)

ISBN 0–19–853969–X

Set by
EXPO Holdings, Malaysia
Printed and bound in Great Britain by
BPCC Hazell Books
Aylesbury, Bucks

Preface

This book is not a biography of August Möbius, nor even a historical study in a straightforward way. Rather, it is a set of essays on topics reflecting upon the contexts, life, work, and influence of a nineteenth-century German academic.

Why Möbius? There were more celebrated mathematicians in nineteenth-century Germany; there were deeper and more famous astronomers; there were university teachers who did more for their institution; there were scholars who participated more fully in the tumultuous events of their day. It is precisely because he was not outstanding in any particular way, but was a serious, competent, professional scholar, that Möbius is such a good mirror of his time. Our aim is as much to illuminate the astronomical and mathematical life of the nineteenth century as to provide a spotlight focusing on only one man; Möbius himself is sometimes of lesser, sometimes of greater, importance in the story.

The six essays in this book cover a considerable range. After the initial chapter, in which John Fauvel sketches the broad outlines of Möbius's background, life, and endeavours, there are two essays analysing the twin academic backgrounds of Möbius's life-work — mathematics and astronomy. First, Gert Schubring analyses the development of the mathematical community in nineteenth-century Germany, and then Allan Chapman describes the revolution that took place in astronomical practice over the period of Möbius's lifetime. There follow two chapters on developments within some of the kinds of mathematics in which Möbius was engaged — geometry, mechanics, and topology. Jeremy Gray analyses what is arguably Möbius's most important mathematical work, his barycentric calculus, and places it in the development both of contemporary mathematics and of Möbius's own work. Then Norman Biggs describes how topological ideas began to acquire their power and versatility during the course of the nineteenth century, and outlines Möbius's role within this process. Finally, Ian Stewart discusses some of the areas of twentieth-century mathematics which have arisen from the concerns of Möbius.

Contents

Adolf Neumann gest.

A. F. Möbius.

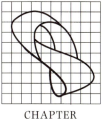

A Saxon mathematician

JOHN FAUVEL

August Ferdinand Möbius was born on 17 November 1790 and died on 26 September 1868. During the course of his lifetime, the pursuit of mathematics in Germany was transformed. In 1790, it would be hard to find one German mathematician of international stature; by the time he died, Germany was the home and training ground of the world's leading mathematicians, and the mathematics researched and taught there spread and came to influence the higher mathematical activity of the rest of the world. We touch on some of the factors effecting this transformation during the course of this book. The changes were not unrelated to the development of the entire German-speaking world over this period, from a gaggle of fairly independent states, through invasions, wars, revolutions, and other tribulations, to an empire united under the political and military might of Prussia.

1790–1816

Möbius was born in Schulpforta, a community in Saxony between Leipzig and Jena — in the centre of Europe, then at a complex and pivotal time in history. To the north, the Prussian state had expanded into the efficient puritanical militaristic bureaucracy bequeathed by Frederick the Great, who had died four years earlier in 1786; to the south, an era was just ending with the death in early 1790 of the enlightened philosopher king Joseph II, the Hapsburg Holy Roman Emperor, based in Vienna; to the west, France, less fortunate in its monarch, was engaged in a revolution whose effects would come to overshadow Europe during Möbius's formative years.

It was very much an age of transition, in the arts and sciences as in politics. In Vienna, for example, Mozart was composing string quartets for the King of Prussia, and had only a year of his short life yet to live. In Bonn, the 20-year-old Beethoven was second viola in the Elector of Cologne's National Theatre. In Weimar, not far from Schulpforta, the 41-year-old Goethe was at the height of his powers, recently back from his formative Italian journey. And in the Gymnasium at Braunschweig, to the north-west, a 13-year-old peasant boy named Carl Friedrich Gauss was eagerly discovering and exploring mathematics.

In France in 1790, the Revolution was under way. It was still in quite a progressive phase, supported with excitement and enthu-

A. F. Möbius (1790–1868).

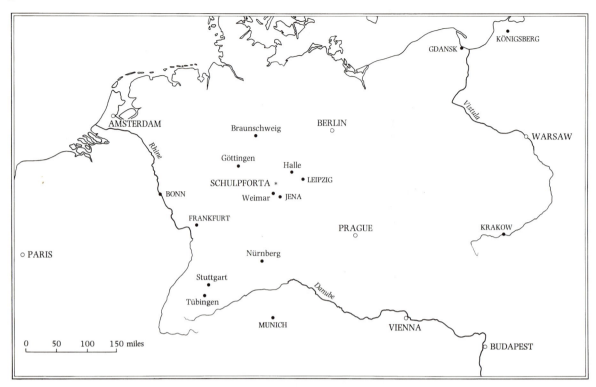

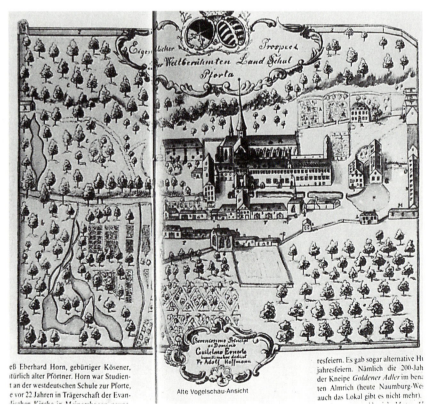

Möbius was born at Schulpforta, a school community where his father taught dancing. Schulpforta was in Saxony throughout Möbius's childhood, but became Prussian in 1815.

eß Eberhard Horn, gebürtiger Kösener, türlich alter Pförtner. Horn war Studien- t an der westdeutschen Schule zur Pforte, e vor 22 Jahren in Trägerschaft der Evan-

Alte Vogelschau-Ansicht

resfeiern. Es gab sogar alternative H jahresfeiern. Nämlich die 200-Jah der Kneipe *Goldener Adler* im ben ten Almrich (heute Naumburg-Wes auch das Lokal gibt es nicht mehr).

siasm by liberal-minded people all over Europe. Many of its innovations are still with us — for example, in May of that year, the French Constituent Assembly passed a law to standardize weights and measures, which turned in due course into the metric system we use today. Only later did the climate become much heavier: Austria and Prussia invaded France within two years; this invasion was repelled, but the King, Louis XVI, was executed and the Revolution turned into the internal terror and external expansionist threat which terrified and appalled its erstwhile sympathizers abroad. By the end of the century, France had moved into a phase of internal stability under Napoleon Bonaparte, although the wars with Austria and Prussia which had started in 1792 had now expanded to include most of Europe.

By 1806, when Möbius was a 16-year-old Schulpforta schoolboy, hostilities between France and Prussia had broken out again, and French troops defeated Prussia and Saxony at the Battle of Jena, just a few miles from Möbius's home. This was a decisive defeat which had a traumatic effect on Prussia and led curiously to a renaissance in German culture, as is discussed in Chapter 2: the shock of defeat led to an upsurge of patriotism and a renewal of education and intellectual life.

The effect on Saxony was different at first, however. Napoleon decided that he needed the ruler of Saxony on his side, because of its sensitive positioning between Prussia and Austria. So he made Saxony

Napoleon accepts the submission of Berlin after destroying the Prussian army at the Battle of Jena, 1806.

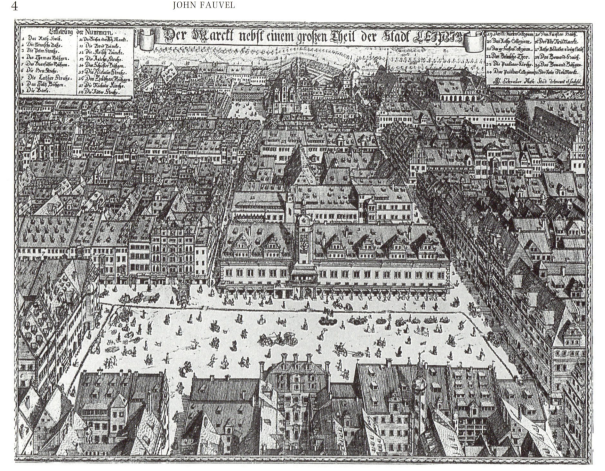

The market square, Leipzig, in an engraving of 1712. The University is at the top of the picture.

a kingdom and formed an alliance, Saxony becoming a sort of French client state for a few years. It was during this period that Möbius entered the University of Leipzig, at the age of 18, in 1809. He first studied law, on family advice, but after a semester he followed his own preferences and studied mathematics, physics, and astronomy. His astronomy teacher, whom he eventually succeeded, was Karl Mollweide, mainly remembered now for a conformal (angle-preserving) map projection he devised, the *Mollweide projection.*

Leipzig is one of the oldest German universities, founded by a migration of German students from the University of Prague in 1409, and at this time had normally around a thousand students and 23 full professors — quite a dominant institution in a town of a little over 30 000 souls. Leipzig University had had some very distinguished teachers and alumni. In the mathematical sciences, several names echo down the centuries: the great fifteenth-century astronomer Regiomontanus was a student in the late 1440s, Copernicus's friend Rheticus taught here in the 1540s (until he had to leave following a homosexual encounter with a student), and Gottfried Wilhelm Leibniz was a student in the 1660s. More recently,

Abraham Kästner taught mathematics at Leipzig from 1739 to 1756, before moving to Göttingen where he was to be an influential teacher of the young Gauss. Remarkably, Kästner influenced, either directly or through his pupils, all three of the mathematicians who are regarded as having discovered non-Euclidean geometry — Gauss, Bolyai, and Lobachevsky. So, in a sense, there was a long-established mathematical and astronomical tradition associated with Leipzig, which Möbius would continue, although it is fair to say that the mathematical education (if any) of most students at any German university before the nineteenth century was on a rather low level.

The Saxon alliance with France went well for a few years, but Napoleon's luck later began to change. In 1813, the French and Saxon troops were defeated at the Battle of Leipzig, Möbius's university town, and the King of Saxony was taken prisoner. This was the crucial battle, the so-called *Battle of the Nations*, which presaged the end of Napoleon's European domination. Eventually, after Napoleon's final downfall in 1815, the Saxon king recovered his kingdom, albeit rather reduced in size.

These historical facts are a significant backdrop to the development of German mathematics, and the life and work of Möbius in particular. Möbius was a Saxon, and this was important to him and to his career. In fact, he had left Saxony in order to visit the University of Göttingen a few months before the Battle of Leipzig, so he was absent when his king was defeated. He had gone to Göttingen to

A contemporary French cartoon satirized the Congress of Vienna, 1815. The King of Saxony, second from right, clutches his crown.

After the Battle of Jena, French troops occupied Leipzig. This contemporary painting shows them searching for contraband.

study theoretical astronomy under the greatest German mathematician of the day, Carl Friedrich Gauss, who was the director of the observatory there. Then he visited Halle, where he studied mathematics under one of the most important mathematicians in Germany at the time, Gauss's teacher Johann Friedrich Pfaff.

In 1814, Möbius learned that Mollweide might be moving to the mathematics chair at Leipzig, which filled him full of hope that he might succeed him in the astronomy position. But when he moved back to work on his *habilitation* — the thesis that would enable him to teach in the University — the outer world of politics caught up with him again, and he was apparently urged to join the army that had been formed by the victorious Prussian-based administration, His comment in a letter to his mother is revealing:

I find it utterly impossible that anyone should think of making me a recruit, me, a fully accredited magister of Leipzig University. This is the most horrible idea I have heard of; and anyone who shall venture, dare, hazard, make bold and have the audacity to propose it will not be safe from my dagger. I do not belong to the Prussians. I am in Saxon service.

We see several fascinating strands in this assertion. It shows partly the horror of any young man trying to avoid military call-up, and partly the slightly pompous pride of the lower-middle-class boy who has made it into a higher social sphere by his own academic efforts, but there is also an evident nationalistic pride in being a Saxon.

And indeed, the story ended happily. Möbius managed to evade being called up, finished his *habilitation* thesis, and sent it off with commendations from Gauss and Pfaff. The King of Saxony managed to salvage half his kingdom from the Congress of Vienna, which was settling the shape of post-Napoleonic Europe, and returned in 1815. And in early 1816, Möbius was appointed Extraordinary Professor of Astronomy at the University of Leipzig, where he stayed for the rest of his life.

August Ferdinand Möbius

1790	Born on 17 November in Schulpforta, Saxony
1809	Student at Leipzig University
1813–4	Travelled to Göttingen (Gauss), Halle (Pfaff)
1815	Wrote Doctoral thesis (*The occultations of fixed stars*) and 'Habilitation' (on *Trigonometrical equations*)
1816	Appointed Extraordinary Professor of Astronomy, Leipzig
1818–21	Leipzig Observatory developed under his supervision
1820	Married: one daughter, two sons
1827	Wrote *The barycentric calculus*
1829	Made Corresponding Member, Berlin Academy of Sciences
1834–6	Wrote popular treatises on *Halley's comet* and the *Principles of astronomy*
1837	Wrote two-volume *Textbook of statics*
1843	Wrote *Celestial mechanics*
1844	Appointed Full Professor in Astronomy, Leipzig
1848	Appointed Director of the Observatory
1855	Wrote *The theory of circular transformations*
1858	Discovered the Möbius band
1868	Died on 26 September in Leipzig

JOHN FAUVEL

1816–1868

The extraordinary professorship held by Möbius was a somewhat lowly form of academic life, meaning that he was entitled to advertise lecture courses for which he might charge a fee. He was not an especially charismatic teacher, and apparently students came to his courses only when he advertised them as free. His progress up the academic ladder was slow. His position was not upgraded to an ordinary chair in astronomy until 1844, and that was only because the University of Jena sought to lure him away. Besides his teaching post, Möbius was appointed Observer at the observatory in 1816. This was his rank for many years. He was finally promoted to Director of the Observatory in 1848.

Leipzig Observatory (1909).

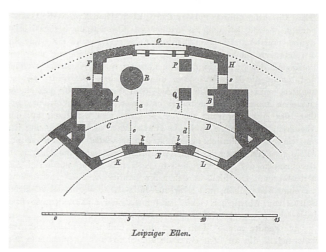

Leipziger Ellen.

Plan of Leipzig Observatory (c. 1820).

August Möbius spent his professional life as an astronomer, although he is mainly remembered now for his mathematical discoveries. Gauss, the greatest mathematician of his age, likewise spent his life as director of an astronomical observatory. This may seem paradoxical to late-twentieth-century eyes, but is partially explained by noting the different social roles of mathematicians and astronomers in early-nineteenth-century Germany. At that time, a mathematician was essentially a poor drudge whose time was spent pumping basic calculations into ill-prepared unmotivated pupils, or if more ambitious was at best an administrator, whereas an astronomer was a scientific professional. It is not a coincidence that two ambitious bright young men from poor backgrounds should choose to enter astronomy as their means of attaining security and respectability. The significance of astronomy and its development in the early nineteenth century is discussed in Chapter 3. Things were about to change in the status of mathematics, but this had not moved very far at the time when Gauss and Möbius were making their career choices.

Mathematical life in Germany and France

Something very remarkable happened in the period coinciding with Möbius's working life. The German-speaking world moved from a fairly basic mathematical level to having possibly the most sophisticated set of mathematicians in the world. In the 1830s, 1840s, and thereafter, there were large numbers of German mathematicians, second to none in competence and productivity. Yet 30 or 40 years earlier, one could have counted them on two fingers. The reasons for this dramatic change are discussed more fully in the next chapter, but here the contrast may be brought out by comparing German with French attitudes towards mathematics during the time of Möbius's childhood and adolescence.

The ideology of the new French Republic from the 1790s onwards was one which promoted mathematics as tremendously significant. The educational system was restructured so as to increase the mathematical training at all levels, and ability at mathematics was a good way for people to get on well. Mathematics was taken very seriously, both intrinsically and as a science in the service of the state. In all Europe, it was France where creative mathematics was primarily happening at the turn of the century. The greatest pure mathematician of the day, the Italian-born Joseph Louis Lagrange, was in Paris throughout the revolutionary years and was a much respected teacher at the new École Polytechnique, and an inspiration to the younger mathematicians of France. There was a galaxy of great mathematicians — Laplace, Monge, Legendre, and so on —

writing textbooks, teaching, and giving lectures, fulfilling commissions for the government, and doing research.

In Germany, there was nothing of this breadth and quality in mathematical activity. Symbolically, it is rather significant that Lagrange had been working in Berlin, at the Berlin Academy of Sciences, but was lured away from Berlin to Paris on Frederick the Great's death in 1786. Indeed, the Prussian kingdom which Frederick left went rather downhill under his successor, until in 1806 there was the tremendous shock of Prussia's defeat by Napoleon's troops at the battle of Jena. This turned out to be rather beneficial, with hindsight, for it led to an upsurge of patriotism and to what was seen as the moral regeneration of the country. Intellectual life was re-evaluated and a flowering of national culture was promoted by educational reform, new institutions, and new social and professional structures. The University of Berlin was founded in 1809, the year Möbius entered Leipzig, and developed during the nineteenth century into the leading institution embodying a new research-oriented professional approach to academic subjects, not least mathematics.

What happened in Germany during those years was not a catching-up with France, although a Parisian training remained the goal of every ambitious German mathematician for the next 20 or 30 years. Rather, it was the growth of a new institutional style, a new way of doing mathematics — something recognizably the forerunner of our university world of professors doing research and teaching and taking graduate seminars. It was, above all, a world of *professional mathematicians* within these institutions, as is discussed further in Chapter 2. The belief was unique to Germany at that time that the function of a professor was not only to pass on knowledge, but to create it too. The kind of mathematics promoted within these structures tended towards what we would call 'pure' mathematics. There was a turning away from the old-style involvement of mathematics with practical and empirical problems; pure research was seen as worthier, and with more spiritual and educational benefits.

Möbius's mathematical work

Although Möbius was an astronomy professor and worked all his life in Saxony, not Prussia, the mood of the new developments impinged upon the whole German-speaking world. In his studies, Möbius was quietly working away at a wide range of mathematical discoveries and creations. His late-nineteenth-century biographer, Richard Baltzer, evocatively described his style of work:

The inspirations for his research he found mostly in the rich well of his original mind. His intuition, the problems he set himself, and the solutions

he found all exhibit something extraordinarily ingenious, something original in an uncontrived way. He worked with unhurried continuity, quietly and in seclusion, in fact almost locked away, until everything had been put into its proper place. Free of haste and far from pomposity and arrogance, he waited for the fruits of his mind to mature. Only after a period did he publish his results in perfect form. ... Everywhere in Möbius's work one can see his endeavour to reach his goals along the shortest path, with the smallest possible amount of machinery and using the most appropriate means.

Our task here is not to summarize and contextualize the full range of Möbius's mathematical work, but simply to explain some of the results for which he is best remembered, especially those to which his name is attached.

His best-known mathematical legacy is, of course, the *Möbius band* or *strip*. This is a one-sided band which can easily be constructed by twisting a strip of paper half a turn before gluing the ends, and arose from his researches in the late 1850s for a Paris Academy prize on the geometric theory of polyhedra. As is explained in Chapter 5, the band was independently discovered a couple of months earlier by Johann Benedict Listing — but it is a time-hallowed mathematical tradition to name things after someone other than the first discoverer! The Möbius band has several interesting properties which can easily be demonstrated using a paper strip and a pair of scissors. The reader may like to try cutting one in half lengthwise, and then cutting another one a third of the way across, asking others to predict what will happen.

The Möbius function

For each natural number n, define

$$\mu(n) \quad = \begin{cases} \quad 1 \quad , \text{ if } n = 1 \\ (-1)^r , \text{ if } n = p_1 p_2 \ldots p_r \; (p_i \text{ distinct}) \\ \quad 0 \quad , \text{ otherwise} \end{cases}$$

The function μ is called the *Möbius function*.
Some values of $\mu(n)$ for small n are as follows:

n	1	2	3	4	5	6	7	8	9	10	11	12
$\mu(n)$	1	–1	–1	0	–1	1	–1	0	0	1	–1	0

Möbius inversion formula. If f is any function, and if

$$F(n) \quad = \sum_{d \mid n} f(d),$$

then, by introducing the Möbius function, we can 'solve' for $f(n)$:

$$f(n) \quad = \sum_{d \mid n} F(d) \, \mu(n/d).$$

For example, if $n = 10$, then

$$F(10) \quad = \quad f(10) + f(5) + f(2) + f(1),$$

and $\quad F(5) = f(5) + f(1), \quad F(2) = f(2) + f(1), \quad F(1) = f(1);$

thus, inverting, we have

$$f(10) \quad = \quad F(10) - F(5) - F(2) + F(1)$$

$$= \quad F(10) \, \mu(1) + F(5) \, \mu(2) + F(2) \, \mu(5) + F(1) \, \mu(10).$$

Möbius's interest in the topic arose from the *reversion of series*:

$$\text{if } F(x) = \sum_{s=1}^{\infty} f(sx)/s^n, \quad \text{then } f(x) = \sum_{s=1}^{\infty} \mu(s) F(sx)/s^n.$$

The Möbius function first appeared in *Ueber eine besondere Art von Umkehrung der Reihen*, 1831.

Möbius transformations

Let C^* be the complex plane, together with the point at infinity. A *Möbius transformation* is a function $f : C^* \to C^*$ of the form

$$f(z) = \frac{az + b}{cz + d}, \quad \text{where } ad \neq bc.$$

Examples of Möbius transformations are:

translations $\quad a = d = 1,\ c = 0,$ so $f(z) = z + b$

scalings/rotations $\quad b = c = 0,\ d = 1,$ so $f(z) = az$

the reciprocal map $\quad a = d = 0,\ b = c,$ so $f(z) = 1/z$

Any Möbius transformation maps circles and lines to circles and lines.
For example,

$$f(z) = \frac{z - i}{z + i}$$

maps the circle $\quad \{\, z : |\, z - i\,| = 1 \,\}\quad$ with centre i and radius 1

to the circle $\quad \{\, z : |\, z + \tfrac{1}{3}\,| = \tfrac{2}{3} \,\}\quad$ with centre $-\tfrac{1}{3}$ and radius $\tfrac{2}{3}$.

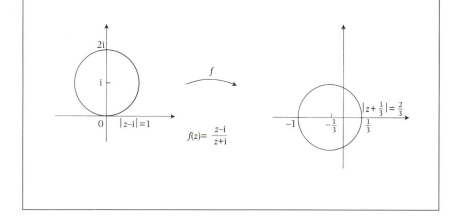

Less well known, but in their contexts very useful, are the *Möbius function* and the *Möbius inversion formula*, the *Möbius transformation*, and the *Möbius net*. These are explained in the accompanying boxes, here and on page 91.

Möbius has also been credited with something he had nothing to do with — namely, the *four colour conjecture*. This conjecture, eventually proved in 1976, states that *any* map (on a plane surface) can be coloured using no more than four colours, in such a way that adjacent areas are coloured differently. For something so easy to state, this conjecture turned out to be remarkably hard to prove. Its

»Der halbe Ton $\frac{135}{128}$ möchte sich noch in Melodieen vorfinden, in welchen
»z. B. die Tonfolge *f fis g* mit den Duraccorden über *f d g* begleitet gedacht wird;
»vorausgesetzt, dass diese Basstöne der *C*-durscala angehören, also

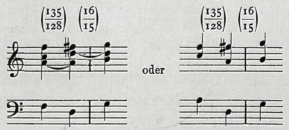

»Das unmelodischeste Intervall, das in Ihrer Tonleiter vorkommt, ist die unver-
»meidliche Quinte $\frac{1024}{675}$, welche zwischen *fis* und *cis* liegt, und hier ist es unmög-
»lich eine Melodie zu ersinnen, die dies Intervall in seiner Reinheit darstellte;
»freilich ist es nur um einen zehntel Ton (um $\frac{2048}{2025}$) grösser als die reine Quinte,
»entspricht also den gewöhnlichen Anforderungen angenäherter Reinheit.

»Die Primzahl 7 haben Sie gewiss mit Recht ausgeschlossen, da sie keine
»anderen melodischen Intervalle liefert als $\frac{8}{7}$, $\frac{7}{6}$, $\frac{15}{14}$ und allenfalls $\frac{7}{5}$.

Möbius was also interested in the theory of musical intervals.

eventual proof took many hundreds of hours of computer time, besides 124 years of mathematicians' work. It was first posed as a question that needed mathematical proof in 1852, by a young British mathematician called Francis Guthrie. In the late nineteenth century, it came to be thought that it was Möbius who had first raised the question, some twelve years earlier than Guthrie, and this attribution was repeated in several popular mathematics books during the twentieth century. What Möbius presented in an 1840 lecture, however, was the following puzzle:

There was once a king with five sons. In his will he stated that after his death the sons should divide the kingdom into five regions in such a way that the boundary of each one should have a frontier line in common with each of the other four. Can the terms of the will be satisfied?

The answer to this problem is *no*, as is not difficult to show by a simple geometrical argument. But unfortunately it has less to do with the four colour conjecture than a superficial glance may suggest. *If* the answer had been yes, *then* the four colour conjecture would be false, but a negative answer proves nothing one way or the other.

A little-known but interesting insight into Möbius's role in inter-disciplinary research is his collaboration with one of his colleagues at Leipzig, the physicist Gustav Theodor Fechner. Fechner was pursuing a programme of mathematizing experimental psychology, working on a new subject or approach which he called *psychophysics*. He was especially concerned with quantifying thresholds of perception, the relationship between external stimuli and mental sensations. Part of Fechner's analysis of sensitivity was based on an underlying probability model; that is, people's judgements (for example, about which of two lines are longer) will naturally vary according to an underlying error function, which has to be taken into account alongside a host of other influencing factors. It was in producing a mathematical model which incorporated inferences from the normal error curve that Möbius was able to help Fechner. This is a small example, but is a helpful reminder of that part of academic life which consists of interacting with colleagues and bringing specialized skills to bear on their projects.

Möbius's popular reputation

Möbius lived a full and academically active life, if not especially eventful externally, up to his death in 1868, which occurred not long after he had celebrated his fiftieth year of teaching at Leipzig. His mathematical influence has lived on in the subjects he investigated (some of which are discussed in Chapters 4 and 5), and the way he investigated them (discussed in Chapter 6). At another level, the Möbius band has become familiar in everyday discourse in much the same way as other mathematical items that everyone has heard of, with varying degree of comprehension, such as Pythagoras's theorem, or squaring the circle. The Möbius band has been portrayed in art by Maurits Escher; in sculpture by Max Bill; in literature in stories by William Upson (*A. Botts and the Moebius strip*), A.J. Deutsch (*A subway named Möbius*), and Martin Gardner (*The no-sided professor*); and on postage stamps (see p.104). In technology,

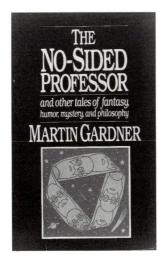

the Möbius band has been the basis of patents for an endless sound record (Lee de Forest, filed 1920), for an abrasive belt (1949), and for a conveyor of hot material (1952). The cultural pervasiveness of the notion of Möbius band is now assured because, rather like some other popular mathematical metaphors, it has begun to be used in all kinds of contexts for which it is thoroughly inappropriate.

> If the world condemns Saddam Hussein for his rape of Kuwait, it can avoid condemning Israel for its rape of Palestine only by resort to an ethical Möbius strip that starts off on one moral plane and ends on another.
>
> Letter in *The Independent*, 17 October 1990
>
> This is a Möbius strip of a novel: a thriller which, halfway through, contorts itself into a novel of ideas, and finally into a romance. In so doing, it snaps its backbone of psychological truth.
>
> Book review in *The Observer*, 13 October 1991

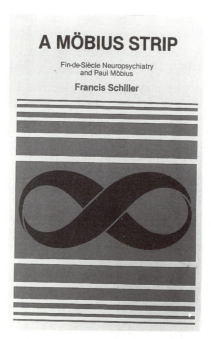

There is sometimes confusion between August Möbius and his grandson Paul, a Leipzig neurologist, who also gave his name to a host of things in his area of research: *Möbius's syndrome*, the *sign of Möbius*, *akinesia algera (Möbius)*, and the like. His work had little in common with his grandfather's, except for one interesting area — or volume, perhaps. On reading Franz Joseph Gall's work on phrenology, Paul Möbius was struck with the fact that the location on the skull

Paul Julius Möbius
(1853–1907).

which Gall had designated the *mathematical organ* — the left fronto-orbital bump — had been an especially prominent feature of his grandfather's head. This led him to research into the shape of mathematicians' heads. He set about collecting them, dead and alive, and wrote a richly illustrated 340-page monograph to demonstrate the correlation between mathematical ability and bumps on the head. Indeed, the renovation at this time of the Leipzig cemetery enabled him to dig up his grandfather's skull to provide the hard evidence. But his hypothesis has not, alas, withstood the test of time nearly as well as his grandfather's mathematics.

Acknowledgement

I am grateful to Costel Harnasz for sharing his knowledge of Paul Möbius.

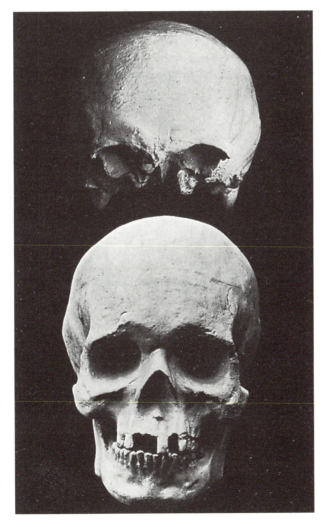

The skulls of A. F. Möbius (above) and L. van Beethoven (below). Paul Möbius's caption to this photograph commented on his grandfather's skull in this way:

'The skull of A. F. Möbius seen from in front. The development of the left corner of his forehead is to be noted; directly behind it one sees the organ of mechanics.'

JOHN FAUVEL

Appendix

> ### Selected list of Möbius's works
>
> 1815   On computing the occultations of fixed stars by means of the planets (Doctoral thesis)
> 1815   An analytical disquisition on certain peculiar properties of trigonometrical equations ('Habilitation')
> 1823   Observations from the Royal University Observatory, Leipzig
> 1827   The barycentric calculus
> 1829   Metrical relations in the area of linear geometry
> 1829   Proof of a new theorem in statics, discovered by Mr Chasles
> 1829   Short description of the main features of a system of glass lenses
> 1831   Development of the conditions of equilibrium between forces acting on a free solid body
> 1833   On a special type of dual proportion between figures in space
> 1834   The true and the apparent orbit of Halley's comet
> 1836   The principles of astronomy
> 1837   On the midpoint of non-parallel forces
> 1837   Textbook of statics
> 1838   On the composition of infinitely small rotations
> 1840   Applications of statics to the theory of geometrical relationships
> 1843   The elements of celestial mechanics
> 1844   Elementary derivation of Newton's laws from Kepler's laws of planetary motion
> 1847   Generalization of Pascal's theorem concerning a hexagon inscribed in a conic section
> 1848   On the form of spherical curves which have no special points
> 1849   On the law of symmetry of crystals and the application of this law to the division of crystals into classes
> 1850   On a proof of the parallelogram law of forces
> 1851   On symmetrical figures
> 1852   Contribution to the theory of the solution of numerical equations
> 1853   On a new relationship between plane figures
> 1853   On the involution of points in a plane
> 1854   Two purely geometrical proofs of Bodenmiller's theorem
> 1855   The theory of circular transformations in a purely geometrical setting
> 1855   On involutions of higher order
> 1856   Theory of collinear involution of pairs of points in a plane or in space
> 1857   On imaginary circles

1858   On conjugate circles
1862   Geometrical development of the properties of an infinitely thin
       bundle of rays
1863   The theory of elementary relationship
1865   On the determination of the volume of a polyhedron

Posthumous works

—   On the theory of polyhedra and elementary relationships
—   Theory of symmetrical figures
—   On an acoustical problem
—   On the calculation of the reserve funds of a life insurance
      company
—   On geometrical addition and multiplication

Further reading

The historical background is well covered in many books — notably, in James J. Sheehan, *German history 1770–1866*, Clarendon Press, Oxford, 1989.

There are not many accounts in English of the life of Möbius. A useful secondary source on his environment and early years is St John Kettle, 'The early life of Moebius, and the world in which he lived it', in J.N. Crossley (ed.), *First Australian conference on the history of mathematics*, Monash University, Clayton, 1981, pp. 145–158.

The status of mathematicians vis-à-vis astronomers in Germany is discussed in Herbert Mehrtens, 'Mathematicians in Germany circa 1800', in H.N. Jahnke and M. Otte (eds.), *Epistemological and social problems of the sciences in the early nineteenth century*, Reidel, Dordrecht, 1981, pp. 401–420. This book contains several other interesting and relevant essays — a good starting point for those interested in exploring this area further.

Fechner's work on psychophysics, and Möbius's contribution, are further discussed in Stephen M. Stigler, *The history of statistics: the measurements of uncertainty before 1900*, Harvard University Press, Cambridge, MA, 1986, pp. 242–254. Also worth reading is J. Ben-David, *The scientist's role in society*, Prentice-Hall, Englewood Cliffs, NJ, 1971. The work and ideas of Paul Möbius may be pursued in Francis Schiller, *A Möbius strip: fin-de-siècle neuropsychiatry and Paul Möbius*, University of California Press, Berkeley, CA, 1982.

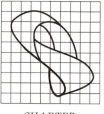

CHAPTER
TWO

The German mathematical community

GERT SCHUBRING

In the nineteenth century, German mathematics reached its peak. During this period, under German influence, mathematics (and especially pure mathematics) became a modern scientific discipline. German mathematicians took over from the French as the international leaders. They carried forward the programme of rigorization which had emerged in France after the revolution.

Between 1800 and 1820 there was only one really important mathematician in Germany — Carl Friedrich Gauss — and young mathematicians used to go to Paris for more advanced studies. But only ten years later, the situation had changed drastically. You can get an impression of this change from the list of important contemporary mathematicians which Georg Simon Ohm drew up in 1838.

Germany		France	Other nations
Gauss	Möbius	Navier	Santini
Bessel	Schweins	Poisson	Plana
Jacobi	Schwerd	Cauchy	Ivory
M. Ohm	v. Staudt	Poinsot	Airy
Dirichlet	Desberger	Poncelet	Quetelet
Steiner	Littrow		
Plücker			

August Leopold Crelle (1780–1855) was a civil engineer whose love of mathematics led to several lasting contributions to the mathematical community. In 1826 he founded the influential mathematical journal still known as *Crelle's Journal*. He promoted the talents and careers of many young German mathematicians, and was employed by the Ministry of Education to advise on the teaching of mathematics in schools and colleges.

In Ohm's judgement, Germany (with 13 names) outstrips all the other countries. Together the others add up to only ten — five French, two Italian, two British and one Belgian; in addition, he mentioned Sturm and Lamé as promising youngsters. Obviously the list is rather subjective, as Ohm includes his brother Martin! It has political overtones too: it was produced for the Bavarian crown prince and includes minor figures from Southern Germany and Austria. Nevertheless, the list does show that in the perception of a German mathematician, the centre of mathematical activity had shifted to Germany — there was no longer any need for Germans to feel mathematically inferior to the French.

Institutional changes

What caused such a decisive change? Naturally, we must consider institutional factors. For example, in one of the larger German states — Prussia — in 1810, the universities and schools were undergoing radical reforms. In the universities the reformers were trying to make the work more scientific and the lectures less dogmatic. The professor's role was redefined, and he was now given the dual role of teaching and research. Traditionally these two functions had been separated, and research had been carried out only in academies.

We see the transition beginning at the University of Göttingen, in the kingdom of Hanover. There, Gauss concentrated almost exclusively on research. He managed to do this by setting high lecture fees to deter students from enrolling for the lectures which he as a professor was supposed to give regularly. Gauss was not keen on introducing students to new fields of research. He regarded teaching as a burden and loaded as much of it as he could onto his colleagues. Nevertheless, there were professors well known for their lectures on elementary mathematics, such as A.G. Kästner and his successor B.F. Thibaut, who attracted a great student audience.

Göttingen University.

11 Waage-Gebäude in Halle

The original building of the University of Halle.

A. G. Kästner (1719–1800).

The protestant universities of Halle and Göttingen were the first to include research as part of the work of the university professor. But what led to this dual role being accepted? How did the professor's research imperative come about, in mathematics and other subjects? It was the result of particular *neohumanist* reforms in Prussia.

In its Prussian form, *neohumanism* meant a plan to reform society through education. The defeat by the French army in 1806 was demoralizing to the Prussians; they lost large sections of their territory. But this proved an incentive for self-examination, and saw the start of an intensive programme of internal reform. It has been called an intellectual revolution, *Geistesrevolution*, rather than a political revolution.

The theme of the reforms was autonomy, *Selbsttätigkeit*, the economic and cultural independence of the individual. The education system became central in society, and teachers became the central agents. Secondary school teachers were given high social prestige — they stood for scientific values and took on the status of scholars.

It was the reform of the philosophical faculties which led to this transformation. These had originally provided a foundation for study in the higher professional faculties, but they now took on the independent role of educating teachers for secondary schools.

Mathematics became a subject in its own right — this was very different from how it had been before. Previously, mathematics had always been taught as part of the philosophical faculty, but neither its professors nor its students had specialized in it. Students used to attend mathematics courses only to complete their liberal education. It was not a subject for several terms of long and specialized study. As a result, the professors who were responsible for teaching such non-specialists were themselves generalists. They taught other subjects as well, or later went on to a professorship in one of the higher faculties.

A telling example of this is Wenzeslaus Karsten, a prominent eighteenth-century German mathematician. He had devoted himself to mathematics from his youth, but because there were no opportunities for a career in mathematics he became a clergyman. He continued to teach himself mathematics and became a tutor, a *Privatdocent*, at the only university in the tiny northern dukedom of Mecklenburg-Schwerin, in Rostock, his local town. (A *Privatdocent*, traditionally the first post in a university career for scientifically qualified persons, was not a salaried position, and the holder relied entirely on the students' lecture fees — he would teach those students who were not taught by the established professors.)

W. J. G. Karsten
(1732–1787).

As there were no mathematics students, Karsten had to accept a chair of philosophy. It was only in 1778, when he was appointed to the chair of mathematics at Halle, that he could devote himself entirely to mathematics. This in itself was a wide-ranging job; it also included all the applications of mathematics such as hydraulics and fortifications.

We can now understand that the Prussian reforms meant a profound change in the previously low status of mathematics. Under these reforms, mathematics rose in status to become one of the three principal subjects taught in secondary schools. These principal subjects now became classical languages, combined history and geography, and mathematics. A mathematical community at the universities and schools emerged, which I shall refer to as a *disciplinary-professional complex*.

Developments in the universities

Because mathematics was a main subject within the neohumanist liberal education, its teachers at secondary schools became a substantial professional group. They gained their internal values and identity from an active interest in current mathematical research, to which they also contributed.

At the same time, the professors developed mathematics into an independent discipline within the universities. The education of teachers was the social function of the discipline. They proposed the idea that pure mathematics provides a philosophical basis for knowledge, which tied in with the current scholarly ideology of the neohumanist universities.

Schulprogrammes of E. E. Kummer, K. Weierstrass, and F. Grashof.

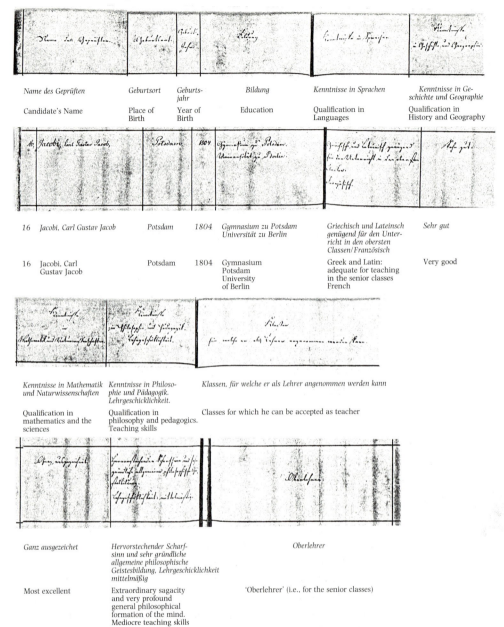

Name des Geprüften	Geburtsort	Geburts-jahr	Bildung	Kenntnisse in Sprachen	Kenntnisse in Ge-schichte und Geographie
Candidate's Name	Place of Birth	Year of Birth	Education	Qualification in Languages	Qualification in History and Geography
16 Jacobi, Carl Gustav Jacob	Potsdam	1804	Gymnasium zu Potsdam Universität zu Berlin	Griechisch und Lateinsch genügend für den Unterricht in den obersten Classen/Französisch	Sehr gut
16 Jacobi, Carl Gustav Jacob	Potsdam	1804	Gymnasium Potsdam University of Berlin	Greek and Latin: adequate for teaching in the senior classes French	Very good

Kenntnisse in Mathematik und Naturwissenschaften	Kenntnisse in Philosophie und Pädagogik. Lehrgeschicklichkeit.	Klassen, für welche er als Lehrer angenommen werden kann
Qualification in mathematics and the sciences	Qualification in philosophy and pedagogics. Teaching skills	Classes for which he can be accepted as teacher
Ganz ausgezeichet	Hervorstechender Scharfsinn und sehr gründliche allgemeine philosophische Geistesbildung, Lehrgeschicklichkeit mittelmäßig	Oberlehrer
Most excellent	Extraordinary sagacity and very profound general philosophical formation of the mind. Mediocre teaching skills	'Oberlehrer' (i.e., for the senior classes)

Report on Jacobi's teacher examination by the Berlin examination board.

The *Gymnasien* — the typical form of secondary schools in Germany — were working on the same social and methodological principles as the universities. It was because of this that there was such unprecedented and fertile ground for the development of mathematics. (In order to enroll at a university, students had first to pass the *Gymnasien* final examinations; later, trial types of secondary schools, called *Realschulen*, grew in strength, and in 1900 they eventually obtained the same right to prepare students for university entrance).

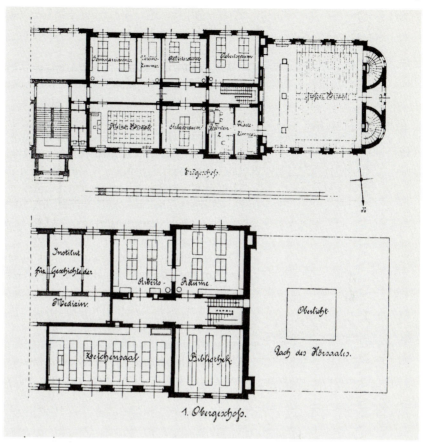

Plan of Leipzig Mathematical Institute.

The professionalization of teachers, and the institutionalization of mathematics as an independent or autonomous discipline, were mutually related. Careers in university and school teaching were also related. Up to the 1860s there were several mathematicians (such as Kummer and Weierstrass) who became university professors after having careers as schoolteachers.

Another development, which also originated in the Prussian universities, became significant for the mathematical community. This was the *seminar*, which provided exercises once a week to complement the normal lecture course, thereby furnishing an alternative to listening passively to lectures. Within the universities, seminars were the only way that students could have scientific contact with their professors, and professors could guide and advise students. The students taking part would have to give a lecture on a topic assigned to them by the professor, and the professor (and to a lesser degree the other participants too) would comment on and criticize the presentation. Furthermore, seminars would be equipped with a small library which would give those taking part access to recent research publications.

The seminar was first established for philology in all Prussian universities after 1810. It proved to be a powerful means, not only of introducing students to recent research, but also of systematically training them in research techniques. This is historically and socially important because people could now become scientists by training, quite apart from any natural genius — it even allowed not particularly talented people to produce significant research results. After some delay, the two other disciplines which were main school subjects (mathematics, and combined history and geography) followed the example of philology and also established seminars.

The first person to apply the seminar model to mathematics was Carl Jacobi, who worked tirelessly to establish and expand the discipline of mathematics. In 1834, together with the theoretical physicist Franz Neumann, he set up the combined mathematics – physics seminar at Königsberg. This seminar eventually produced a large number of highly qualified teachers, as well as famous mathematicians. The seminar system was introduced in Halle in 1839. The other universities did not follow until the 1860s, when the number of students generally began to increase. These seminars were the seeds of the later Mathematical Institutes.

The development of mathematics as a *disciplinary–professional complex* led to its being studied as a pure, and not as an applied, subject. There are several aspects to this.

First, at the school level, mathematics was taught in a particular cultural context in which formal mental training was valued. This stimulated teachers to strive for clarity in the basics, for a logical

C. G. J. Jacobi (1804–1851).

G. P. L. Dirichlet
(1805–1859).

order, and for purity in the use of methods, so fitting in with the view that mathematics is a training for the mind. This meant that they did not stress a particular practical goal and then work out which methods would best arrive at this goal. They tried to follow just one method in order to train methodical thinking.

At the university level, a discipline could best survive, and even expand, if it was accepted as an autonomous discipline, rather than an auxiliary one. If we remember the aim of making subjects autonomous, we can now understand why Jacobi was so militant about rejecting externally defined values such as usefulness, and why he emphasized the internal values of the discipline. His proud report to his brother of how he shocked the 1842 conference of natural scientists at Manchester is well known:

There I had the courage to make the valid point that it is the glory of science to be of no use. This caused a vehement shaking of heads.

In his inaugural lecture as ordinary professor at Königsberg in 1832, Jacobi criticized the French mathematicians for putting too much stress on *applied* mathematics, and for mixing up the true and the incidental causes of progress in science:

We are unhappy that most French geometers who originate from the school of the famous Laplace have presently fallen into this error. While they seek to obtain the only salvation for mathematics in physical problems, they desert that true and natural path of the discipline, which ... has brought the analytical art to the importance which it now enjoys.

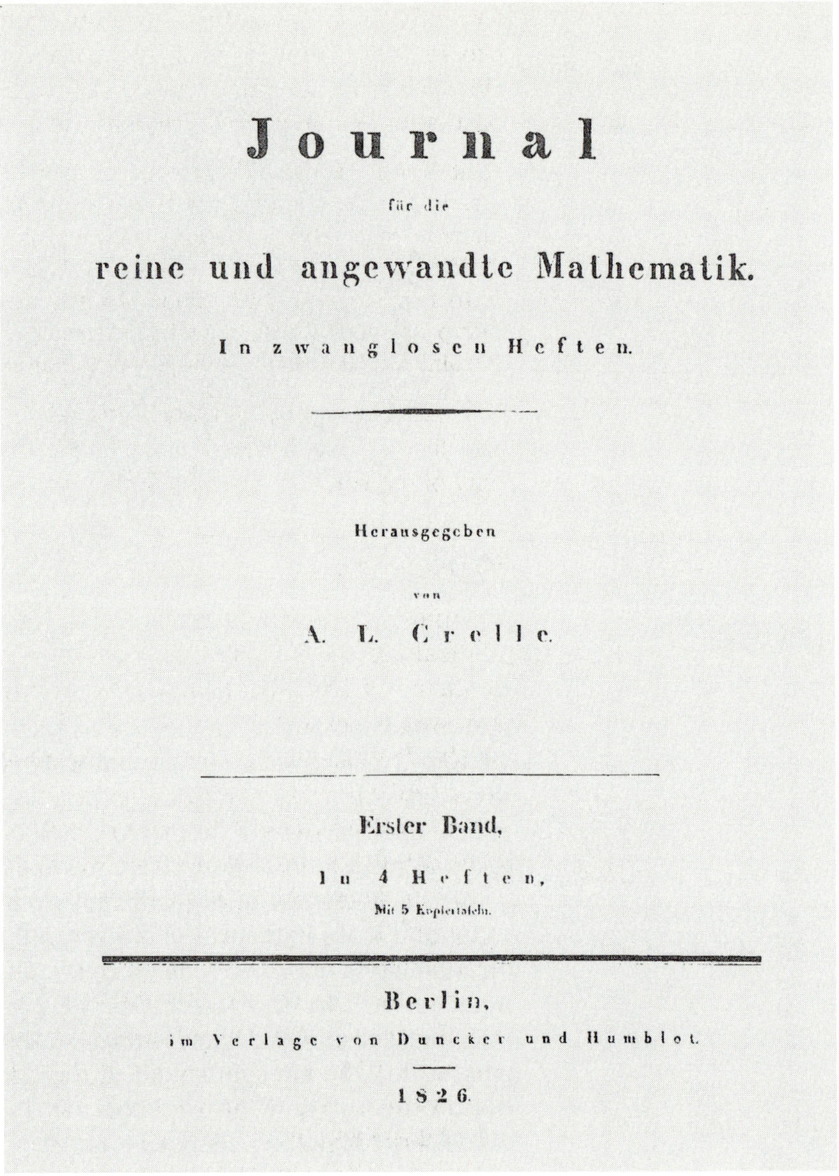

Journal

für die

reine und angewandte Mathematik.

In zwanglosen Heften.

——————

Herausgegeben

von

A. L. Crelle.

———————

Erster Band,

In 4 Heften,

Mit 5 Kupfertafeln.

————————————

Berlin,

im Verlage von Duncker und Humblot.

——

1826.

Jacobi added therefore:

In this way it is not so much pure mathematics, but its application to physical problems, that suffers.

Someone who repeatedly pointed this out was August Leopold Crelle. As consultant in mathematics to the Prussian Ministry of Education from 1828 onwards, he had quite an influence on appointments and promotions in universities. Crelle's understanding of the relation between pure and applied mathematics was as a hierarchy. Pure mathematics provided the methodological basis for meaningful and coherent applications. This explains why he argued fervently for there to be more pure mathematics when he gave his expert advice on a Polytechnic Institute that was planned for Berlin. He declared that pure mathematics constitutes a system with a particular degree of connectedness, and continued:

So it is also important that pure mathematics should be explained in the first instance without regard for its applications, and without its being interrupted by them. It should develop purely from within itself and for itself, for only in this way can it be free to move and evolve in all directions. ... In teaching the applications of mathematics, it is results in particular that people look for. They will be extremely easy to find for the person who is trained in the science itself, and who has adopted its spirit.

One must bear in mind that rational mechanics, to which Dirichlet and Jacobi made important contributions, was regarded by Crelle as part of pure mathematics.

Publications

Journals played an important part in the growth of the mathematical community. We find that the journals which were published confirm our idea about the way in which the mathematical community was composed.

The *Journal für die reine und angewandte Mathematik* (Journal for pure and applied mathematics) was established in 1826 by Crelle and promoted by the Prussian ministry. It became the first permanent scholarly journal for the emerging mathematical community. Many of the decisive papers by Jacobi, Dirichlet, Möbius, and others were published here. Until the 1860s, a considerable number of the German contributors were teachers.

Soon there were two other journals set up for teachers. The *Archiv der Mathematik und Physik* (Archive of mathematics and physics) was established in 1841 by J.A. Grunert, and was important for promoting rigour and clarifying problems about basics. The third, *Zeitschrift für Mathematik und Physik* (Journal for mathematics and physics), was set up in 1856 by Oscar Schloemilch, and made research results more accessible.

Until recently, historians have largely overlooked another peculiar type of German publication — the *Schulprogramme*. Every year one member of the teaching staff of each secondary school was required to publish a dissertation. This system was first introduced in 1824 in Prussia in order to promote scientific activity among teachers, and we find a number of highly interesting papers on problems of rigour and fundamentals: the theory of parallel lines, negative numbers, basic notions of the infinitesimal calculus, and so on. These rigour- and system-oriented teachers also made a substantial contribution to the presentation of the elements of mathematics. It is clear that later mathematicians who were taught by them were profoundly influenced by this style of mathematics.

The *Schulprogramme* reveals another important feature. Publications on arithmetic, algebra, and analysis, and those on geometry, were of about the same quantity. So we need to revise the generally held view that during the nineteenth century there was just one trend, towards *arithmetization*.

In fact, at this time there was also a revival in geometry. This revival was brought about by the stress on purity of method. We find that, from the 1820s onwards, there was a growing move in Prussia away from algebraic geometry, because it used mixed methods. As a result, we find a return to the purely geometrical methods for conic sections by geometrical loci, together with a growing acceptance of the new synthetic geometry, and in particular the methods of Jacob Steiner which were especially influential in schools.

The trend towards geometry

There was yet another factor which reinforced geometry against the trend towards arithmetization.

Although I have mainly been concerned with Prussia, other parts of Germany were rather different — if anything, the emphasis was even more strongly on geometry in these states, and this also had its influence. In the other German states, there had been no comparable social and educational reforms. No deep institutional reforms took place in the universities and schools, and mathematics maintained a rather marginal or auxiliary status.

A characteristic example is Bavaria. In classical secondary schools, mathematics was a subordinate subject whose purpose was to pass on some traditional and popular notions of geometry and astronomy. Mathematics held a stronger, but still auxiliary, position in the schools of agriculture and commerce. These schools, along with the central polytechnic of Munich, fell under the jurisdiction not of the Ministry of Education, but of the Ministry of Commerce.

It is remarkable that in this different cultural and institutional context we find yet another style of mathematics. Here we find no praise of pure mathematics and no outstanding work on algebraic analysis. It was geometry that was cultivated, culminating in Christian von Staudt's research on synthetic and projective geometry. Originating from the polytechnic context — that is, the entire system of polytechnic schools (such as the Munich Polytechnic) which existed throughout Bavaria — there was increasingly an accent on *descriptive geometry*. That descriptive geometry should be fostered in Bavaria and Austria is even more remarkable as there was almost no interest in this in northern Germany.

This period of the emergence of German mathematics, which one could call the *revolutionary period,* was characterized by the unity of the discipline *and* the profession in Prussia. But during what we might call the normal phase, from about 1870, this unity fell apart. There was a split between a community of university mathematicians who specialized more and more in sub-theories, and a community of school-based mathematicians who no longer felt obliged to do any research.

In 1869 Friedrich Richelot, Jacobi's successor in Königsberg, demanded that differentiation be made in the final examinations between future mathematical researchers and future mathematics teachers. The Ministry granted this demand, which marks the splitting of the discipline from the profession.

A separation between pure and applied mathematics had also come about. Crelle's original conception of a hierarchical ranking, in which pure mathematics provided the methodological basis for applied mathematics, had now turned into a separation of disciplines. No longer were mathematicians able to carry out work in both directions, as Gauss or Jacobi or Plücker had done in the past.

So what had been a united discipline was now divided. This division was later to lead Felix Klein to look for ways to reunify mathematics, and to find again the link between theoretical pure research and the stimulus that comes from the applications of mathematics.

Acknowledgement

I am grateful to Johannes Karsten, Wismar, for sending me the autobiographical report of his ancestor Wenzeslaus Karsten, written in 1766.

Further reading

There are a number of accounts of the material in this chapter. Among those in English are R.S. Turner, 'The growth of professorial research in Prussia', *Historical studies in the physical sciences* **3** (1971), 137–182; and D. Rowe, 'Klein, Hilbert and the Göttingen tradition' in K. Olesko (ed.), *Science in Germany,* Osiris (2nd series) **5** (1989), 186–213.

Among several articles by the author are 'The conception of pure mathematics as an instrument in the professionalization of mathematics' in H. Mehrtens *et al.* (eds), *Social history of nineteenth century mathematics,* Birkhäuser, Basel, 1981, pp. 111–134; and 'Pure and applied mathematics in divergent institutional settings in Germany: the role and impact of Felix Klein' in D. Rowe and J. McCleary (eds), *History of modern mathematics,* Vol. II, Academic Press, New York, 1989, pp. 171–220.

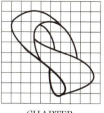

The astronomical revolution

ALLAN CHAPMAN

The 78 years of Möbius's life witnessed a fundamental shift in the status of Germany as a 'great power' in the realm of the intellect. While that confederation of states collectively called *Germany* had produced many of the great theological intellects of the sixteenth century to pioneer the Reformation, the region had been devastated by the Thirty Years War which raged in the seventeenth century. Relatively speaking, early- and mid-eighteenth-century Germany was still a European backwater, a constant theatre of war, and generally off the route of the European 'Grand Tour'; it was regarded as provincial in most things other than music and perhaps certain branches of literature.

Germany and the Enlightenment

But its numerous states, Prince-Bishoprics, and Electorates contained a large number of academic establishments, even if many of them were obscure in terms of the wider world. By the late eighteenth century, however, Germany was displaying many portents of things to come — with Kant, Goethe, Schiller, and Beethoven in philosophy and the arts, and Mayer, Humboldt, Olbers, and Schröter in the sciences.

Although the German states were not politically unified, and still enjoyed varying degrees of self-government, the German language and the academic universality of Latin made it easy for professors and other talented men to accept posts in states of which they were not native. The Napoleonic wars further ravaged the country, but even while the conflict with France was being fought, a remarkable intellectual life was in the process of development. Although this renaissance was taking place in virtually all aspects of intellectual life, from the linguistic analysis of ancient texts to experimental chemistry, the scientific or critical methodological approach was coming to lie at the heart of them all. In some respects, this intellectual revival contained a patriotic dimension, expressing a new national identity as the life of the mind of a region which was emerging out of internal chaos and foreign aggression.

The universities in the German intellectual renaissance

The revival of institutional higher education in Germany was heralded in 1809, when Friedrich Wilhelm, the King of Prussia,

The 10-inch heliometer by Adolf Repsold for the Radcliffe Observatory, Oxford, 1848. This magnificent refractor, like all heliometers, had a divided object glass, the segment alignments of which were governed by fine screw-rods from the eyepiece end. It was mounted on a delicately poised clock-driven 'German' equatorial mount. Precision tracking of stars was crucial when measuring angles to a fraction of an arc second.

The University of Berlin, now known as Humboldt University.

founded the University of Berlin. Because of its political significance and Royal foundation, vast sums of public money were expended on it as it consciously 'head-hunted' the best professors and students. Kaiser Wilhelm was also to establish the Rhenish University of Bonn in 1818, while Munich University was re-founded, along with several others that had been forced out of existence by the wars.

Almost from the start, these well-endowed new universities enjoyed excellent facilities, with libraries, astronomical observatories, and (soon after) laboratories, as obvious parts of their equipment. Perhaps for the first time in the history of higher education, physical science was included as a natural component of the curriculum, to be studied alongside classics and theology.

The growing wealth and princely patronage of Berlin University meant that many academics would win distinction in other academic institutions before being lured to Berlin. In particular, Franz Encke (after 1825) and Johann Galle spent the main part of their working lives at the Berlin Observatory, although others refused its blandishments and remained loyal to their own foundations.

Wilhelm Bessel, perhaps the greatest observational astronomer of his time, refused to leave Königsberg, although Königsberg was another Prussian Royal foundation. Carl Friedrich Gauss stayed at Göttingen, while Möbius himself, as a loyal Saxon, turned down tempting positions at Greifswald and Dorpat Observatories (before Berlin Observatory achieved its real eminence) to remain for his entire working life at Leipzig.

Higher education in Britain

The German universities of the early nineteenth century differed considerably from those of England, although perhaps less so from those of Scotland. Prior to the founding of the University of London in 1828, Oxford and Cambridge were the only English universities, and catered for a relatively narrow clientele of young Anglican gentlemen. In many respects, they were finishing schools where young men from the wealthy classes were given an intellectual and social polish before being sent out into the world. Their primary purpose was to *teach*, rather than to *conduct research*. Whatever research was conducted in Oxford and Cambridge was usually unofficial and, while regarded as wholly meritorious, was not an expected part of a professor's job.

The four Scottish universities, being much cheaper to attend than Oxford and Cambridge and open to non-Anglicans, attracted a more diverse body of students, although the *instruction* of students was still the predominant requirement of the professoriate, and not original research.

The primacy of original research

One of the most dynamic features of the new German universities was their stress upon research as well as teaching. The requirement of a doctor's degree by all those holding a professorship meant that all German academics possessed some research experience. It was the German universities of this period that invented the modern Ph.D. degree, along with the research departments to prepare graduate students to take it. In all branches of learning, from theology, via history, to mathematics and medicine, the investigative spirit of critical research was fostered, as higher education was seen to be synonymous with pushing back the frontiers of knowledge.

The great German universities — Berlin, Heidelberg, Göttingen, Leipzig, Munich, and others — attracted students from all over the world, as English, French, Italian, and American scholars came to attend their graduate schools. These men, deeply impressed by the German model, went back to reshape their home universities. Following the opening-up of the West, the flood of new universities in the United States often copied the German foundations with Ph.D.

graduate schools, observatories, and scientific research stations. The new English civic universities of Manchester, Leeds, and Birmingham also borrowed crucial German elements of organization, even if their architecture was Oxbridge inspired. German, moreover, was coming to oust Latin as the international academic language — a far cry from its provincial obscurity in the previous century — and by 1860, any young man who planned a scientific career would learn German as a matter of course.

The German scientific renaissance

The physical sciences played a major part in this German academic renaissance and, for a variety of reasons, enjoyed a peculiarly high prominence. This prominence derived in part from a new intellectual authority which science was seen as possessing, as an instrument of 'demonstrable truth', but another stemmed from its high cost. Astronomical observatories in particular required expensive pieces of capital equipment; even a wealthy amateur like Johann Schröter, of Lilienthal near Bremen, was eventually obliged to sell his observatory to King George III of England (and still ruler of Hanover) in return for the continued use of the instruments during his lifetime, prior to their being turned over to the University of Bremen.

Heinrich Olbers
(1758–1840).

Comets attracted a lot of attention, and astronomers who studied them were highly regarded. This picture shows the great comet of 1811, studied by Olbers.

Möbius also contributed to the popularization of comets. This title page comes from his 1834 book on Halley's comet.

Heinrich Olbers: Germany's Grand Amateur astronomer

Perhaps the only great German astronomer of this period who neither held a university appointment nor ended up in financial difficulties was Heinrich Olbers of Bremen. His independence derived in part from the substantial fortune which he made from medical practice — he was one of the leading ophthalmologists of his day — as well as from the specialized character of his astronomical research. Olbers was perhaps the foremost comet expert of his generation (1758–1840); indeed, he discovered several. Although he possessed an excellent collection of instruments to pursue this work, he needed fewer massively expensive pieces than he would have done had he been working in other branches of the science. These instruments were within the financial range of a well-to-do professional man, as was his magnificent library. Olbers's library was the most complete cometographic library then extant, and when he died in 1840, it was purchased for the Russian observatory at Pulkowa, near St. Petersburg.

Observatories as official astronomical institutions

A fundamental component in the rising prominence of German astronomical research was the newly founded, or re-founded, institutional observatories of the post-Napoleonic era. In 1813, Friedrich Wilhelm of Prussia established the observatory of Königsberg in the extreme north-east of his kingdom, not far from the Russian border. This was to become the research home of Friedrich Wilhelm Bessel (1784–1846). Altona Observatory near Hamburg was completed in 1823, with Schumacher as its director. The Berlin Observatory had been in existence since 1705, although it was re-founded and re-equipped with the finest instruments between 1832 and 1835, while the King of Prussia sanctioned the establishment of a new observatory for the University of Bonn in 1836.

Since 1790, Möbius's University of Leipzig had possessed an astronomical observatory, to which he had been appointed Extra-ordinary Professor and Observer (a fairly junior post) in 1816. In 1848, Möbius became Director of this observatory, which was eventually to be remodelled in the 1860s. But if Möbius's promotion up the professorial hierarchy was slow, we must remember that this derived in part from certain special circumstances. On the one hand, his loyalty to the University of Leipzig and dislike of travel made him reluctant to go elsewhere. On the other hand, his real interests lay in mathematics, rather than in astronomy as such, although he failed in his attempt to obtain the Leipzig mathematical chair. In consequence, he was something of a mathematical fish in observational astronomical water: an incongruity which cannot have helped his career.

Yet ironically, two of the most illustrious observatories of the early nineteenth century were not in Germany but in Russia, although they were German equipped and directed. In 1809, the University of Dorpat, in the Russian province of Livonia, maintained a small observatory. In 1813, the Directorship of this re-equipped institution was given to the north-German born and educated Wilhelm Struve (1793–1864). He was to become the first of a distinguished dynasty of astronomers of that name extending down to the present day. In 1839, Wilhelm Struve moved from Dorpat to take charge of the new observatory at Pulkowa. Although a Russian institution under the auspices of the St. Petersburg Academy of Sciences, Pulkowa (like Dorpat) was German in its instrumentation, direction, and style of research.

Astronomy in France and Britain

France had enjoyed a brilliant astronomical reputation in the eighteenth century, culminating perhaps in the celestial mechanics of Laplace in the 1790s, but the Paris Observatory (not to mention

Wilhelm Struve (1793–1864), Director of Dorpat, and later Pulkowa, Observatories.

Dorpat Observatory in Estonia. Built in 1809–10, the central structure originally carried a dome.

those in provincial France) was inferior to German institutions by 1835. Symptomatic of this problem was Urbain Le Verrier's solicitation of German (and English) assistance to search for his *planet X*, Neptune, in 1846; this resulted in the first sighting of the new planet by the Berlin Observatory on 23 September 1846.

By 1845, the only country which maintained a significant astronomical tradition that was not Germanic in conception was Great Britain, although even there, German achievements and practices were admired and sometimes adopted. Adverse comparisons with German work, moreover, had played a part in moving the Royal Society and Board of Visitors to overhaul the working procedures of the Greenwich Royal Observatory in the 1820s and 1830s. George Biddell Airy, who became Astronomer Royal in 1835, was a Cambridge man to his fingertips, but had been brought up with his eyes fixed on the achievements of contemporary continental science. During his undergraduate days in the late 1810s, these had consisted of French mathematical analysis and (by 1835) German mathematics and astronomy.

The Royal Observatory in Greenwich, *c.* 1870.

The new programmes of systematic observation and error analysis which Airy instituted at Greenwich after 1835 were reminiscent of those at Dorpat and Königsberg, although Airy imparted to Greenwich important elements of his own which kept it quintessentially English. One of these elements was his dependence on instruments of English manufacture, for England was the only nation in Europe whose astronomical instrument makers were still capable of competing with the Germans in 1835. The second was the sheer efficiency of the Royal Observatory, which maintained 24-hour monitoring of the heavens, and annually published its observations for the benefit of the world of learning. Airy had supplied astronomical observations to Le Verrier in Paris, just as he had done to Neptune's English co-discoverer John Couch Adams, although he saw no obligation to search for other people's planets, be they French or English.

While Airy and his friends in the Royal Astronomical Society were among the few major astronomers of the day to operate outside the German orbit, he nonetheless felt at home in Germany, and regularly acted as host to visiting German men of science. I am not aware that Airy ever met Möbius on his many visits to Germany, although it is likely. He certainly met, and was on very friendly terms with, a host of Möbius's colleagues, including Schumacher, Encke, D'Arrest (Möbius's son-in-law), Hansen, and Struve, as well as receiving German, French, and Russian scientific decorations to honour his achievements at Greenwich.

Yet the rapid advances in German astronomical instrument making were already making themselves felt in England. In the midst of a nasty legal wrangle with the English instrument-making firm of Troughton and Simms in 1832, Sir James South visited the Dorpat Observatory to seek ways of improving on Troughton and Simms's imperfect mount for his new 11.75-inch-aperture refractor. South went to Dorpat to examine Joseph Fraunhofer's superb 9.6-inch-aperture equatorially mounted refractor, installed in 1824 and still the largest and finest refractor in the world. By 1842, indeed, the Oxford University Observatory had placed an order with Adolf Repsold of Hamburg for a large new heliometer. The heliometer was a precision refracting telescope with a divided object glass, used for making critical measurements between astronomical bodies. It was a class of instrument which the Germans made very much their own.

German astronomical priorities

Several issues dominated astronomy intellectually during Möbius's lifetime. Broadly speaking, they can be broken down into four main areas:

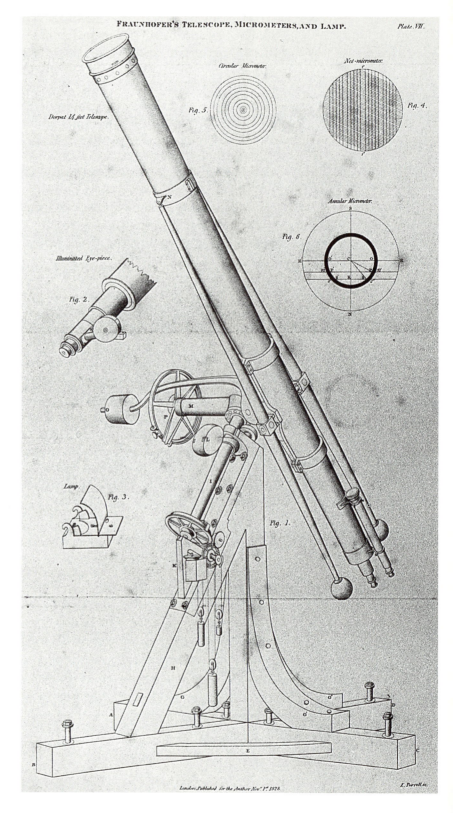

FRAUNHOFER'S TELESCOPE, MICROMETERS, AND LAMP.   Plate. VII.

Circular Micrometer.   Fig. 5.

Net-micrometer.   Fig. 4.

Dorpat 14 feet Telescope.

Annular Micrometer.   Fig. 6.

Illuminated Eye-piece.   Fig. 2.

Lamp.   Fig. 3.

Fig. 1.

London, Published for the Author Nov.r 1.st 1828.

E. Purcell sc.

(*a*) the *measurement and quantification* of the Solar System in accordance with Newton's Laws;

(*b*) *astrometry*, or the critical measurement of star positions in order to compile maps and tables from which a wide variety of investigations could proceed;

(*c*) *cosmology and astrophysics*, building upon the work of Sir William Herschel in England, whereby new techniques of 'gaging' the structure of the universe were devised — but unlike Herschel's purely visual techniques, the new science of astrophysics (based as it was on new optical and chemical discoveries) developed methods whereby the physical composition of deep space objects could be investigated;

(d) the *physical investigation* of the Sun, Moon, planets, and comets by superior optical resolution to map and examine their surfaces.

The measurement of the Solar System

The intellectual coherence of eighteenth-century astronomy rested on the *laws of gravitation*, first propounded by Sir Isaac Newton in 1687. Newton's laws provided both the theoretical and computational basis for a whole new understanding of planetary dynamics, especially after Leibniz's independent invention of the calculus improved upon and simplified Newton's 'fluxions' technique for calculating gravitational attractions between orbiting bodies.

It had been the elaborate convolutions of the lunar orbit which had most engaged the attentions of eighteenth-century astronomers, for it was realized that if the Moon's motions could be predicted with critical accuracy on the basis of Newtonian techniques, then they could be used to find a ship's longitude at sea. The British Government had offered a substantial cash reward to astronomers and mechanicians who could perfect the longitude technique (be it by 'lunars' or chronometer), and part of this prize had been paid to the widow of Tobias Mayer of Göttingen. It had been Mayer's tables of the lunar motions, based on observations made with a quadrant by the Englishman John Bird, which first allowed the computation of a ship's position using the Moon.

Bode's law and the asteroids

One of the most remarkable achievements of eighteenth-century astronomy had been the establishment of accurate dimensions for the Solar System, as planetary distances, masses, and velocities were found to be in agreement with Newton's predictions. The distances of the planets accorded with a precise scale, proceeding from the Sun outwards. There seemed, however, to be a break in this sequence,

Joseph Fraunhofer's magnificent 9.6-inch-aperture refractor for the Dorpat Observatory, 1824. Note also Fraunhofer's 'German' mount, by which the weight of the telescope was delicately counterpoised upon the equatorial axis to facilitate the most accurate tracking of objects. Fraunhofer's prototype design of large refractor and German mount was supplied to Königsberg, Berlin, and many of the great European observatories.

between Mars and Jupiter, and

it was, therefore, thrown out by the late Professor Bode of Berlin, as a possible surmise, that a planet might exist between Mars and Jupiter.

Johann Bode (1747–1826), Director of the Berlin Observatory, had constructed an empirical mathematical model in 1772 (which was based upon a refinement of a suggestion by Titius of Wittenburg), in which the sequence of numbers $\{4 + 3.2^{n-2}, n \geq 2\}$ was found to correspond, with an extraordinary degree of accuracy, to the known distances of the planets from the Sun. His sequence ran thus:

4	7	10	16	28	52	100
Mercury	Venus	Earth	Mars	?	Jupiter	Saturn

This 'law' of Bode's seemed to display an added cogency after 1781, when William Herschel's newly-discovered outer planet Uranus was found to occupy a position very close to the next predicted place in the sequence. Uranus fell at 192 on the scale, whereas Bode's next number would have been 196.

The problem, needless to say, centred on the number 28, which represented the unoccupied place in the sequence and which prompted Bode to posit the existence of a hitherto undiscovered planet. It was obvious that, if such a planet existed, it must be a telescopic object which had eluded naked-eye observers since antiquity. In consequence, a small committee of astronomers met, largely under the aegis of Johann Schröter of Lilienthal, with the intention of 'policing' the zodiac plane of the sky in which all the known planets moved and where an unknown one might be found.

Yet Schröter, Olbers, von Zach, and the other 'celestial policemen' working in the last months of 1800 found nothing by systematic search. It was in the extreme south of Europe, well away from the main centres of astronomical investigation, that the crucial chance discovery was made. One might say that the astronomy of the 'minor planets', or asteroids, could not have been more a part of the nineteenth century, for it was on the night of 1 January 1801, the first night of the new century, that Giuseppe Piazzi discovered the first of such bodies from his observatory at Palermo, Sicily. The Palermo Observatory was very well suited for this type of work, for its extreme southerly location in Europe at 38° north meant that the zodiac was higher up in a clearer sky than ever it could have been in England, France or Germany. It was thus ideally placed to search for dim planetary bodies. It was also excellently equipped: amongst other instruments, it contained Jesse Ramsden's celebrated 5-foot circle of 1789, the first large circular instrument in the world to be used in fundamental research.

It came to be realized that this newly discovered object — which Piazzi named *Ceres* after the classical corn goddess of Sicily —

Azimuth circle by Jesse Ramsden for the Palermo Observatory, 1789. This instrument of revolutionary design could observe horizontal and vertical angles on its circular (not quadrantal) scales.

occupied a place between Mars and Jupiter, and before it moved out of a favourable location for observation, several angular measurements of its position in relation to the fixed stars were secured.

It was at this juncture that the German astronomers successfully entered into the quest for minor planets. Piazzi's relatively small number of measurements of the motion of Ceres were not adequate to compute a proper orbit using the existing mathematical techniques. It was the 23-year-old Carl Gauss in Göttingen who applied a

Carl Friedrich Gauss
(1777–1855) on the terrace
of Göttingen Observatory.
Behind him stands a portable
heliometer, probably by
Repsold.

mathematical technique of his own devising to Piazzi's values and
computed an orbit. Exactly one year after Piazzi's original discovery,
on 1 January 1802, Olbers in Bremen re-found Ceres on the
strength of Gauss's predicted place. Not only did this event herald
brilliant prospects for Gauss, but it laid the foundations of a lifelong
friendship between him and Olbers.

It was Olbers who discovered the second asteroid, *Pallas*, in March
1802, while Karl Harding at Schröter's nearby Lilienthal Observatory discovered a third, named *Juno*, in 1804. Olbers was impressed
by the orbital similarity of these three asteroids, and posited that
they probably derived from a disintegrated planet. This opinion was
strengthened when, watching the three known asteroids move closely
towards each other in the spring of 1807, Olbers discovered a fourth,
which he called *Vesta*.

Minor planet astronomy rapidly became a German specialism. It
represented a dramatic confluence between prediction, observation,
and orbital computation, which vindicated Newton's laws by demon-
strating their application to hitherto undiscovered bodies. Although
Bode's law was entirely empirical, and derived from nothing more

than a juggling of numbers, it was remarkable that the newly discovered asteroids were found to occupy the vacant '28 place' on his scale, which may once have been the actual orbit of a disintegrated planet. We still possess no theoretical basis to explain Bode's law. Once discovered, however, the asteroids were found to follow the individual orbits predicted for them by Newtonian dynamics.

Möbius and planetary orbits

Möbius had studied under Gauss at Göttingen in 1813, and it is interesting to note that two of his earliest published papers dealt with the orbits of Pallas and Juno. Möbius's astronomical interests were concerned with celestial mechanics, and in 1816 he wrote a paper devoted to an analysis of the azimuth variation in the position of the fixed stars. He also paid attention, as in his Juno and Pallas papers, to the techniques of orbital computation for newly discovered asteroids and comets. The reappearance of Halley's comet in 1835 spurred Möbius to produce two popular treatises which analysed the comet's orbit and the wider laws of astronomy.

We must not forget that Halley's comet was seen as possessing considerable scientific significance when it returned in 1835. This was, after all, only the second of the comet's predicted returns, and its orbital characteristics were still based on measurements made by astronomers in 1758–9. The 1835 reappearance enabled astronomers to secure much more reliable measurements than had been possible 76 years before, and use them to apply rigorous tests upon Newton's laws and perfect further their knowledge of solar system dynamics.

Möbius was not only interested in writing pure mathematical treatises for fellow scientists, but also desired to reach the already growing community of amateur astronomers. His *Mechanik des Himmels* (1843) set out to present the basic (and not so basic) information of celestial mechanics to the amateur.

Astrometry

Astrometry, or the measurement of the angular separations between celestial bodies, is one of the oldest concerns of astronomy. Since the earliest star maps of Hipparchus in the second century BC, astronomers had been trying to measure the precise distances between the stars, although it was not until the late seventeenth century that rapid progress was made. This was brought home to astronomers by the presence of a major cosmological question: if the Earth actually goes around the Sun in accordance with the Copernican theory, then some stars should display a slight shift, or *parallax*, as the Earth moves across the base line of its orbit every six months.

The stellar parallax

It was the actual *measuring* of this tiny angle which really posed the problem. By 1700, astronomers realized that, even if the stars were distributed evenly throughout space, the nearest of them must still be vastly remote by Solar System standards, and hence must display a very small angle. Eighteenth-century astronomers had tried to determine the parallaxes by measuring the positions of bright stars against dim ones, working on the assumption that if God had made all the stars of equal luminosity (in the same way that he had made all oak trees and cows of the same generic size), then the bright ones should be nearest to the Earth and hence display bigger parallax angles.

Yet all attempts to measure these parallaxes and establish such distances had failed. By 1728, the English astronomer James Bradley argued that the stellar parallax must be less than half of one second of arc, for his instrument was capable of measuring down to that angle and had detected nothing.

All of this early parallax work was performed in England. What made it possible was the development of measuring instruments fitted with telescopic sights and precision micrometers. It had been in England, after about 1670, that major innovations in instrument making had taken place, with Tompion, Sharp, Graham, Bird, and Ramsden, so that by 1800 it was generally acknowledged that London was the home of the finest astronomical measuring instrument makers in the world. It was through a rapidly developing instrumental tradition in the eighteenth century that a new high-level astrometry became possible.

All of the major German observatories were involved in astrometrical work during this period, as were those of England, France, and elsewhere. The more routine aspects of astrometry consisted of making accurate positional measurements of all the stars in the sky, down to a given magnitude, to form maps and tables of ever-increasing accuracy. The most exhaustive celestial cartographic work undertaken in Germany was probably at the Bonn Observatory, where Friedrich Argelander established the coordinates of 324 189 stars down to the ninth magnitude, between 1837 and 1862. These celestial cartographic measurements were made with large graduated circular instruments capable of denoting angles to a fraction of a single second of arc. Once compiled, these tables could be used to provide stellar marker points against which to measure the positions of asteroids, comets, and planets.

They could furthermore be used to measure and identify eligible stars that might display the coveted parallax, and thereby extend a measuring rod out into space, so that precise dimensions could be

Friedrich W.A. Argelander
(1799–1875), Director of the
Bonn Observatory.

ascribed to the stellar universe in the same way as eighteenth-century astronomers had quantified the Solar System. This activity formed the real cutting edge of German astronomy in Möbius's time. Before one could hope to measure a parallax, however, one had to decide on a particular star whose seasonal motion *might* be detectable. Choosing an eligible star from the thousands visible through even a modest telescope was not easy, although some well-defined guidelines had been laid down by 1830. For one thing, it was now realized that brightness was not necessarily the key factor, for as Olbers had already pointed out, the presence of obscuring interstellar dust destroyed any guarantee that dim stars were necessarily more remote than bright ones.

William Herschel, working in England in the early 1780s, had attempted to measure parallax by a novel method. By 1781, he had recorded the locations of numerous double star pairs visible in the northern hemisphere, and had selected from them what he believed to be suitable 'line of sight' doubles that were not *gravitationally*

William Herschel (1738–1822). After serving as a musician in the Hanoverian Footguards, he settled in England. He laid the foundations of deep-space astronomy and recognized the important 'light-gathering' power of large reflecting telescopes for this work.

related to each other, but only appeared close from our direction of sight in space. It is true that, at this stage, Herschel selected stars where a bright and a dim star were found close together. At this date, neither he nor any other astronomer had really recognized that space contained opaque obscuring dust, and considered star brightness to be intrinsic and affected only by distance.

Herschel's technique was to view the chosen pair of stars in the same telescope field, and with an accurate micrometer, measure any apparent movement which took place between them over six months. It soon became clear, however, that no parallax was measurable between any of them, although Herschel's systematic study of double stars initiated an area of research which would become fundamental to nineteenth-century astrometry.

The study of double stars would eventually provide a key to the stellar parallax, although based upon different theoretical criteria from those pursued by Herschel. But they would also lead to the demonstration of Newtonian gravitation in stellar space, as physical pairs of stars, or *binaries*, were shown to exert an influence upon each other. In this respect, another example of the true universality of Newtonian gravitation could be demonstrated, as the same mass and distance laws that bind the Earth to the Moon and the planets to the Sun were now shown to operate with similar effect many light years away between incandescent stellar bodies. The man who was to succeed in both of these fields of investigation was Friedrich Wilhelm Bessel (1784–1846) of the Königsberg Observatory in Prussia.

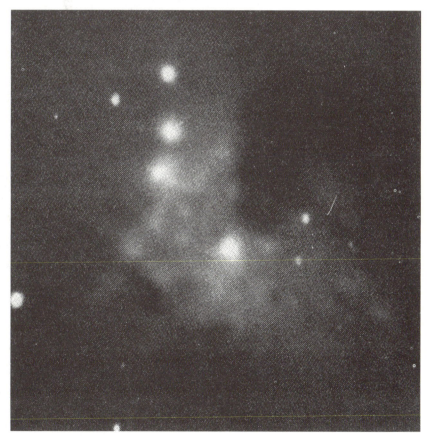

An early photograph, taken in 1857, of a double star.

Bessel and 61 Cygni

Bessel ranks amongst the greatest of German astronomers, and was certainly the man who laid the foundations of astrometry in his lifetime. He combined in one person those two great divisions of

F.W. Bessel (1784–1846).
Note the romantic
'Beethovenian' pose in which
early-nineteenth-century
Germans often depicted their
intellectual heroes.

astronomical creativity which in most cases are found only in separate individuals — he was both a superlative theoretician *and* a meticulous practical observer. Bessel possessed the scientific imagination to see along which avenue the theoretical solution to the parallax problem might be found, and then capped it by being an observer and mechanician capable of finding the solution experimentally. I would suggest that only William Herschel was equally gifted in both theory and practice.

As a youth, Bessel had evinced no outstanding talents for anything. Only after being apprenticed to a merchant in a Bremen office did he discover a flair for numbers by way of accountancy; this interest led him to navigation, astronomy, and cosmology. In 1804, his work was brought to the attention of Heinrich Olbers who resided in Bremen and was instrumental in obtaining for him a post at Schröter's nearby observatory at Lilienthal in 1806. His ascent now proved meteoric, and in 1810, Bessel became Director of the Prussian Royal Observatory at Königsberg (although its instrumentation was not complete until 1813), which was to provide the theatre of his extraordinary achievements over the next 36 years. Nor must one neglect Olbers in Bessel's career, for it had been through this Bremen physician and 'amateur' astronomer's extraordinary generosity of spirit and good connections that both Bessel and Gauss were launched internationally: neither man forgot his debt of gratitude.

The earliest photograph of a solar eclipse was taken from the Königsberg Observatory, in 1851.

The foundations of Bessel's astrometrical researches were historical as well as contemporary. He recognized that the published Greenwich Royal Observatory star positions had been of an acceptable critical accuracy since 1750, when the Astronomers Royal, James Bradley (1742–62) and Nevil Maskelyne (1764–1811), first began to apply a wide variety of delicate corrections to their observations. Bessel began to compare their measures for certain important stars with those of his own day, so that he could obtain critical values for their positions, to see if any motion was detectable.

Bessel approached the stellar parallax problem from a different stance to any of his predecessors, and in this respect, the stars in the Greenwich observations were of the greatest importance. Instead of attempting to measure the assumed parallaxes of bright stars against dim ones, he searched for stars of any brightness which displayed large proper motions. The proper motions of a select number of stars had first been detected by Edmond Halley in 1718, and were the independent motions which certain stars displayed when their positions were measured over long periods of time. Bessel argued that the proper motion is probably a line of sight effect produced by the shift in position of the entire Solar System with respect to a particular star as the Sun moves through space. He argued that, if this were the case (as indeed it is), then stars which display the largest proper motions are probably closest to the Solar System and hence are likely to display the biggest parallaxes as well.

The largest proper motion known in the 1830s was that of the relatively dim *61 Cygni,* or the 61st star in Cygnus the Swan in Flamsteed's 1725 catalogue. Its proper motion was 5.2" arc per year, added to which it was a double star with a slightly dimmer companion.

By the autumn of 1838, after 18 months of observation, Bessel had secured enough measurements of 61 Cygni, its companion, and other stars of lesser proper motion in the same field, to compute a parallax of $\alpha = 0.314"$ arc; this is very close to the value of $\alpha = 0.292"$ arc obtained from modern photographic measurements for the same star. These measurements, secured with a superb Fraunhofer heliometer (a type of instrument which will be discussed presently), made it possible to calculate a distance in the region of 10 light years for 61 Cygni.

At last, a stellar distance had been measured, although almost at Bessel's heels, Wilhelm Struve at Dorpat announced a parallax for Vega in 1839 while, at the Cape of Good Hope Observatory, the British astronomer Thomas Henderson simultaneously announced a parallax for the southern hemisphere star α Centauri, which we now know to be the star closest to our own Solar System.

The successful announcement of three parallaxes within a few months admirably demonstrates the way in which the international spread of astronomical data, plus the availability of instruments made in accordance with new standards of accuracy, made it suddenly possible to cross the ancient limiting thresholds of physical investigation.

Bessel's observing techniques

But Bessel's astrometric work set a standard by itself. As already pointed out, this lay in part in comparing star positions which he had measured himself with the best measures of the same objects from the previous century, although this was not the whole story. Bessel was one of the first astronomers to realize that, before a positional observation could be fully relied upon, one must have quantitative knowledge of every possible error that might enter into the finished result. He came to use Bradley's and Maskelyne's eighteenth-century Greenwich observations because these two astronomers were the first to provide exhaustive analyses of their own instrumental errors, along with temperature and pressure of the atmosphere through which the measurements had been made. By eliminating *all* sources of error — optical, mechanical, and meteorological — Bessel was able to obtain astrometrical results of astonishing delicacy from which a great deal of new data could be extracted.

A splendid example of this approach came when Bessel examined Sirius and Procyon, which had been fundamental stars in Maskelyne's catalogue. Comparing his own observations of these stars with Maskelyne's, he found regular variations in their proper motions which he correctly ascribed to the presence of 'dark companion' stars, the gravitational attractions of which caused a wobbling motion on their brighter parent stars. Both Sirius and Procyon were later found to possess companion stars.

Much of this astrometric work, performed in the major observatories of the day, came to focus upon the study of double stars, the usefulness of which, as 'closed systems', facilitated the extraction of all sorts of data. Parallaxes, proper motions, and the possible demonstration of gravity acting between binary pairs (or triples) were all fields which made double stars so useful to the astronomers of the early nineteenth century.

In addition to their astrometric potentialities, double and compound stars formed ideal test objects for the comparison of telescope resolving powers. In 1862, for instance, the American optician Alvan Clark, when testing a new 18-inch diameter object glass on Sirius, discovered its 'B' companion — the smaller companion star, first predicted by Bessel from the characteristics of Sirius's irregular proper motion! In this way, the power of mathematical analysis could receive dramatic substantiation from the advancing cutting edge of practical optics, to make the theoretical and practical departments of astronomy work hand in glove. It also provided an excellent example of the presence of Newtonian gravitation operating between stellar bodies.

Franz Encke and the Berlin School of Astrometry

If Bessel's Königsberg Observatory pioneered refined astrometry in Germany, it was in the other Prussian foundation at Berlin where it was to proceed on a relentless, almost industrial, basis. Indeed, Bessel had been offered the Directorship of the Prussian Academy of Sciences Observatory in Berlin in 1825, following the retirement of Bode, but had turned it down, fearing that it would bring with it an inundation of administrative duties. Staying at Königsberg, he recommended the 34-year-old Prussian Johann Franz Encke for the post. Encke possessed the exact combination of talents that was needed: a former pupil of Gauss at Göttingen, already a distinguished astronomer, Director of the Seeberg Observatory, and calculator of the orbit of the comet which was later to bear his name. He was, moreover, a thorough and forceful man, an excellent organizer, teacher, and inspirer, who succeeded in turning the Berlin Observatory into a training ground for a whole generation of European astronomers. His effectiveness became even greater after 1835, when

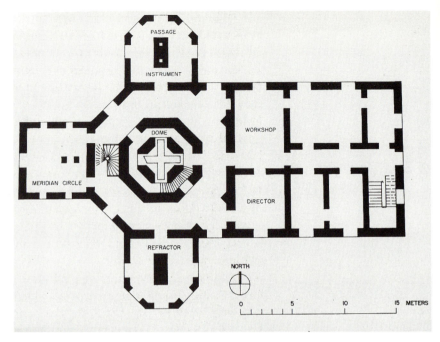

PASSAGE

INSTRUMENT

WORKSHOP

DOME

MERIDIAN CIRCLE

DIRECTOR

REFRACTOR

NORTH

0   5   10   15   METERS

The Berlin Observatory at the end of the nineteenth century.

funds were granted for the rebuilding and equipping of the observatory with the finest instruments.

In many respects, Encke was similar to his younger contemporary, the English Astronomer Royal G. B. Airy, who set about the wholesale revision of the Greenwich Observatory in 1835 at the same age at which Encke had first taken charge of Berlin ten years before. In addition to similarities of temperament, gifts, and outlook, both men recognized that relentless astrometric revision of stellar coordinates lay at the heart of precision astronomy. If Encke was an inspired teacher of young academic astronomers, Airy was no less adept in 'drilling' his non-graduate assistants in the meticulous performance of routine astronomical and mathematical tasks.

Perhaps where Airy and Encke differed most markedly, however, was in their respective approaches to the *skill* of astronomical observation. Encke, like Bessel, Olbers, and most of the other leading German astronomers of the day, enjoyed practical observation and performed it with great dexterity. Airy, on the other hand, openly admitted that it was a wholly unintellectual activity, and that 'an idiot, with a few days' practice, may observe very well'. But Airy was chronically astigmatic and suffered from generally poor eyesight, which may have deprived him of a pleasure and caused him to underrate an activity enjoyed by most of his Continental colleagues with sharper vision. In spite of this conspicuous disparity of attitude towards observation, Airy and Encke enjoyed an amicable and cooperative relationship over thirty years.

When Encke's assistants, Johann Galle and Heinrich D'Arrest, succeeded in locating the position of the new planet Neptune on 23 September 1846, at the coordinates computed by Le Verrier in Paris, the speed of identification was made possible by the fortunate completion by the Berlin astronomers of a new and as yet unpublished star chart of the precise region where the planet was expected to be.

It also says something about Airy's attitude towards the international character of scientific discovery that he was willing to acknowledge 'Le Verrier's planet' and its German discovery, without feeling that his own country and university man, John Couch Adams of Cambridge, had a prior claim, or that Greenwich *could* have initiated the search and beaten Berlin. Yet Adams had never taken the trouble to communicate his *complete* results, whereas Le Verrier had, while the Berlin Observatory was an academic foundation intended for research, as opposed to a relatively 'routine' government institution existing to supply coordinates to the navy, as Greenwich was.

It says something for the international circles in which Airy moved, that he first received news of the Berlin discovery when dining with Professor Hansen on a visit to Gotha in 1846. Unfortunately for Airy, however, many of his more patriotic colleagues back in London were incensed by his willingness to acknowledge a Franco-German discovery, while at the same time failing to recognize a less precise prior claim by an Englishman.

In these displays of French chauvinism, the cartoonist disparages British claims to have discovered the position of Neptune.

Cosmology and astrophysics

During Möbius's life, the principal concerns of astronomers internationally were directed towards the delicate angular measurements discussed so far. On the other hand, the cosmological dimension was never very far away, as scientists speculated not so much about the distances of the stars, but rather the physical structure of the universe of which they were a part.

But the great limitation which applied to this aspect of astronomy before about 1850 was the lack of solid measurable data upon which to base scientific *conclusions*, as opposed to *speculations*. William

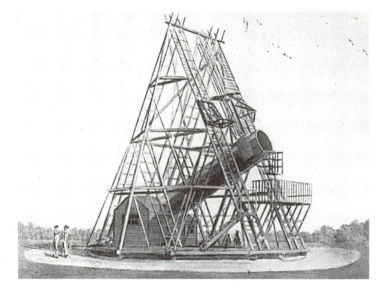

(*a*) William Herschel's 48-inch 40-foot focus reflecting telescope. The 4-foot-diameter mirror, which Herschel built in England in 1789 to attempt to resolve nebulae, was part of the largest telescope in the world for over 50 years.

(*b*) John Herschel's photograph of his father's telescope. Dating from 1839, it is one of the earliest photographs ever taken.

Herschel, in the 1780s and 1790s, had broken down the first barriers when he came to realize that reflecting telescopes of large aperture could collect more optical data than small refractors, and thereby penetrate further into space. He had shown that the Milky Way forms a flat plane, that inter-stellar nebulosity or 'shining fluid' exists, and he suggested that the dynamics of star-cluster formation take place under the influence of gravity.

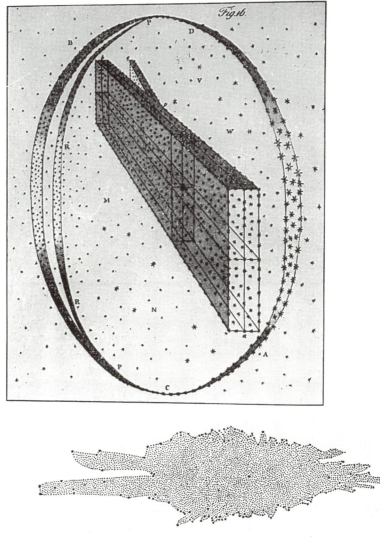

(*a*) Herschel's construction of the shape of the galaxy from the distribution of visible stars. The apparent 'split' of the galaxy is now explained, not in terms of a breaking-away of stars, but their being obscured by opaque inter-stellar dust.

(*b*) The cross-section of the galaxy with its 'split'. Herschel has incorrectly located the solar system at the centre.

But Herschel had sometimes speculated beyond his actual evidence, interspersing his celestial models with philosophical and aesthetic components drawn from a wider scheme of ideas. For instance, he assumed that all stars are of the same generic size and intrinsic brightness, and that they crash together to form dense

Lord Rosse's 6-foot-diameter
reflecting telescope at Birr
Castle, Ireland, in 1845.

globular clusters, while the whole universe is beset by two forces:
condensation under the force of gravity, and the recycling of
resulting debris to form new stars in accordance with a steady state
model.

By the time of his death in 1822, Herschel had taken cosmology
as far as it could go with the purely visual observing instruments
and techniques then available. It is true that astronomers puzzled
endlessly beyond the horizon of measurable perception, while some,
like Lord Rosse in Ireland, tried to push perception further back by
building yet bigger speculum mirror telescopes of the Herschel type.
Rosse inquired into whether the nebulae are resolvable into separate
physical components, while Olbers had posed his celebrated paradox:
*if the stars extend forever through our space, why does their combined
light at midnight not outshine the Sun at noon?* The solution is now
recognized as lying partly in the absorption of much starlight by
opaque inter-stellar dust, although we now know that much is lost
through the increasing red-shift of progressively distant stars.

But the discovery of the concept of the red shift, in the early
twentieth century, was itself a product of that chain of break-
throughs in nineteenth-century physics and chemistry which gave
cosmology a new yardstick and made possible the new science of
astrophysics — the *spectroscope*. Although the original discovery that

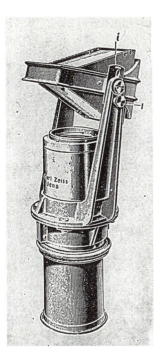

Large glass prism placed in front of a telescope object glass. It was with such an apparatus that Fraunhofer obtained the spectrum of Sirius and other bright stars.

light can be analysed into its component colours by passing it through a prism had first been recognized by Newton in 1666, and improved by Francis Wollaston in 1802, the creation of the spectroscope as a tool of physical enquiry was a product of German optical research and technology.

Joseph Fraunhofer's optical researches

In 1814, Joseph Fraunhofer, already established as the leading optician and telescope maker in Germany, encountered a phenomenon which emerged as a side-effect of other researches. In his attempt to find a source of pure monochrome light with which he could analyse the refractive indices of optical glasses without interference from chromatic disturbances, he observed black lines crossing the solar spectrum along with the colours. Fraunhofer was not the first to see these lines, for they had been noticed by the Englishman Francis Wollaston in 1802, when he viewed a solar spectrum through a *slit*, as opposed to the classical Newtonian pinhole. Wollaston had found that, in addition to the classical Newtonian colours, the slit produced a few dark lines which crossed the spectrum, although he did not pursue the matter.

When Fraunhofer rediscovered these lines, he did so with an apparatus that was superior to Wollaston's. He placed a fine flint-glass prism in front of the object glass of a small theodolite telescope and found that it produced hundreds of thick and thin black lines crossing the resulting solar spectrum. These lines were found to be permanent (as opposed to a product of the glass or air), and when he put a larger prism in front of the object glass of a 4.5-inch-aperture refracting telescope and examined the sunlight reflected from the Moon and planets, he found that they stayed in exactly the same place. On the other hand, when he looked at Sirius and other bright stars, the lines were found to occupy *different* places.

Fraunhofer's 1814–15 spectrum-map of the Sun, showing the black lines that bear his name, the characteristic colours, and the letters with which he designated them.

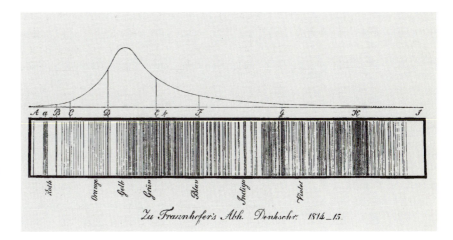

Joseph Fraunhofer
(1787–1826).

Fraunhofer 'mapped' the positions of over 574 of the Sun's black lines, giving the characteristic letters *A, B, C, D, ..., H* to the most conspicuous of them, by which they remain known to this day. For testing optical glass and finished lenses, these black 'Fraunhofer' lines provided useful neutral benchmarks, and were to become one of the major tools in the science of astrophysics by indicating chemical and physical data about incandescent sources.

The spectroscopic identification of elements

While not primarily concerned with the physics of the Sun and stars himself, Fraunhofer had demonstrated experimentally that the dark lines are produced by the light-emitting bodies at source, and are not secondary effects introduced by the terrestrial atmosphere. Even more remarkable, he found that the positions of the two dark *D* lines in the solar spectrum correspond with the same luminous yellow lines that are produced by the light of an incandescent sodium flame. This discovery provided the first indication that it might be possible to detect chemical substances in the Sun and stars.

Yet Fraunhofer did not make the crucial connection between black 'absorption' and coloured 'emission' lines to identify sodium in the solar spectrum. That work was performed independently by Foucault in Paris and Miller in London during the 1840s. While burning sodium salts in the flame of an electric arc, Foucault obtained the bright yellow lines. When sunlight was permitted through the same incandescent flame, however, it 'intensified' the *D* lines in the absorption spectra. From this he concluded that incandescent elements can both *emit* coloured spectra and *absorb* (and hence blacken) the spectral light passed through it.

Spectroscopic apparatus for experimentally determining the coincidence of solar and metallic element lines in the laboratory. The equatorially mounted and clock-driven plane mirror on the window ledge directs a ray of sunlight into the bench spectroscope. The carbon arc, placed between the two condensers in the path of the sunlight, is then switched on. When metallic elements are burned in its flame, they appear as coloured 'emission' spectra in place of their characteristically black 'absorption' lines in the solar spectrum. In this way, it became possible for Kirchhoff, Bunsen, and Huggins to identify incandescent elements in the Sun's light.

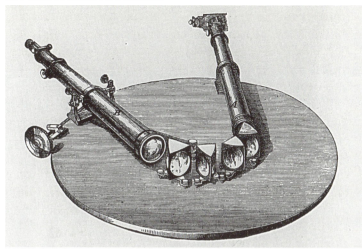

Spectroscope with four prisms, of the type used in laboratory analysis of light.

A great deal of work on the nature of spectra was performed in Britain by Sir David Brewster and by Professors Forbes, Miller, and Stokes. They were interested in why these *emission* and *absorption* spectra (as they were later called) are produced, and how they are related. It was Bunsen and Kirchhoff in 1859 who provided the unifying explanations. Kirchhoff discovered that the spectral characteristics of absorption and emission lines are temperature-related, and that in the laboratory he could get different effects with different flame temperatures. He therefore drew the conclusion that elements in the glowing incandescent mass of the Sun are 'surrounded by a gaseous atmosphere of somewhat lower temperature...[and]...from the occupancy of these [D] lines, the presence of sodium in the atmosphere of the Sun may therefore be concluded'.

Over the next two decades, physicists in the laboratory tabulated and measured the exact spectral places of most of the known elements, while astronomers in the observatory came to identify them in the spectra of incandescent heavenly bodies. William Huggins and W. A. Miller in England became the main pioneers of this technique of spectral analysis and chemical identification of stars. This process advanced at a remarkable speed in the late nineteenth century, especially with the development of photographic techniques of recording delicate spectral lines, and indicates the way in which a breakthrough in instrument technology can generate a whole range of discoveries in pure science.

The physical astronomy of the planets

While the astrometrical and astrophysical branches of astronomy were pushing back the frontiers of man's understanding of the structure of the universe, a great deal of work was being accomplished in the physical examination of the planetary bodies.

Schröter's 27-foot reflector at the Lilenthal Observatory, in 1793. The long wooden tube and optics are carried around and elevated in the altazimuth by the elaborate wheeled wooden structure.

The work of Johann Schröter

The leading figure in late-eighteenth- and early-nineteenth-century observational astronomy was Johann Schröter (1745–1816), a lawyer, magistrate, and 'grand amateur' of Lilienthal, near Bremen, in Hanover. Like his fellow Hanoverian and correspondent, William Herschel, Schröter's initial interests in astronomy were triggered by an early interest in music. He purchased instruments from Herschel (then resident in England), including a 15-inch-aperture and 27-foot-focus reflecting telescope, and was one of the few major German astronomers to do fundamental work with mirror telescopes. The Germans, one might say, were a 'refractor people', and Schröter's use of the reflector may have had something to do with the fact that his observing career was virtually over before Joseph Fraunhofer began to revolutionize the size, resolving power, and image quality of large refractors.

Schröter's observatory at Lilienthal was a place of fundamental research in the physical astronomy of the Moon and planets for some thirty years, and would probably have been so for longer had it not been seriously damaged by Napoleon's troops. It was also the place where, on 21 September 1800, six astronomers met with the intention of searching for a planet between Mars and Jupiter in the wake of Bode's suggestion, discussed above.

Schröter was not primarily concerned with the astrometrical branches of astronomy, but with the prolonged, systematic, and detailed observation of Solar System bodies. It was Schröter who examined the lunar surface for signs of change and, in the absence of any photographic yardsticks of comparison, invented a scale of visual light intensity with which to compare the respective luminosities of topographical features. He also studied the structure of sunspots and, in 1796, published a treatise on the planet Venus. Schröter observed the generally featureless surface of Venus over three decades, during which time he felt convinced that he had detected mountain tops poking through the cloud cover.

Using these supposed features, along with minute changes in the shading of the clouds, Schröter believed that he had estimated the Venusian period of axial rotation. He also observed Mercury, as well as reporting lines and geographical features upon the surface of Mars. Unfortunately, we now know that his values for the rotation periods of Mercury and Venus were quite erroneous, along with his explanation of various Martian topographical features. Even so, we must not forget that Schröter was the first astronomer to pioneer the long-term topographical study of the Moon and planets.

Schröter was the first to classify many features on the lunar landscape and made excellent, if often unpolished, drawings of lunar features, as well as inventing descriptive terms still in use today. It was Schröter, for instance, who first used the word *Rille* (from the German for *groove*) to describe the characteristic cracks found in some regions of the lunar surface. But lacking the permanent record of the photographic plate and the analytical power of the spectroscope, he was entirely dependent upon fleeting visual images seen by eye at the telescope.

Schröter's Lilienthal institution was perhaps Germany's first observatory devoted to pure research, as opposed to forming an adjunct to university teaching. In this respect, it constituted a focal point for the astronomical community in North Germany. In addition to his initiation of asteroid research and planetary studies, his

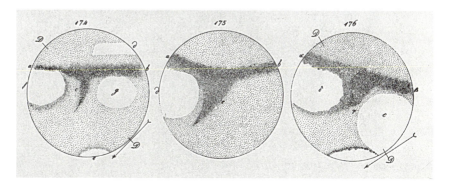

Three of Johann Schröter's drawings of Mars made in 1800 with a 9.3-inch-aperture reflector.

observatory formed a major training ground and point of contact for what was going on in German research. It was with Schröter's aid that Karl Harding, Bessel, and the selenographer J. F. J. Schmidt were launched, for he was a significant patron of rising talent.

German lunar observers

Johann Friedrich Mädler (1794–1874) was perhaps the outstanding physical observational astronomer of the next generation. Employed initially by the Berlin banker and amateur astronomer Wilhelm Beer, he used a 95 mm Fraunhofer refractor, in conjunction with a good micrometer, to construct the first really accurate topographical chart of the lunar surface, published in 1837. In terms of lunar topography, he took up from where Schröter had left off, searching for evidence of change in the dead landscape and studying the relationship between craters, 'seas', rilles, and mountains. Mädler became a professor at Dorpat in 1840, although he returned from Russia to Germany in 1865.

Yet because they searched for signs of subtle change in the absence of an impartial recording medium, and lacked the solid substantive techniques of astrometry, these physical astronomers of the planets were often in danger of an over-imaginative interpretation of their data. Perhaps the most imaginative of them was Franz von Gruithusen who came up with some startling 'finds' in the 1830s. He interpreted the ashen light of Venus, in which the dark body of the planet glows dimly when only the crescent is fully illuminated, as caused by public bonfires lit by the Venusians to celebrate periodic festivals. Von Gruithusen's over-fertile imagination also led him to interpret a collection of lunar geological features as the outlines of a city built by Moon dwellers.

The German scientific instrument trade

None of the areas of astronomical inquiry discussed so far, especially celestial mechanics, astrometry, and astrophysics, would have been possible without a fundamental revolution in the design and use of instruments during this period. By 1800, astronomy had become the most highly developed of the sciences. At a time when medicine and biology were still discussed in terms of vague imponderable forces, astronomy already possessed a firm foundation of measurement and mathematical prediction upon which to ground further research. This foundation was made possible by the development of two classes of instrument — angular instruments for measuring celestial position, and optical instruments which not only gave images of increasingly clear resolution, but made possible the new analytical techniques of spectroscopy and photography. Although England dominated both of these branches of instrument making in 1800, Germany was soon to begin a craft tradition of its own, which came to provide the measurement technology upon which a new standard of astronomical discovery could be built.

Reichenbach's circles

Between 1791 and 1793, Georg Friedrich von Reichenbach, an engineering officer in the Bavarian army, visited England to meet its leading men of science and study the fruits of its Industrial Revolution. In particular, he met Jesse Ramsden, the leading circle-divider and precision instrument maker of the day, and was deeply impressed by the quality of English scientific instruments, which were then superior to any of German manufacture. At this time, England enjoyed a very close political and military relationship with Germany. The English monarchy still possessed territories in Hanover, and both countries shared a common enemy in France — especially, post-Revolutionary and Napoleonic France.

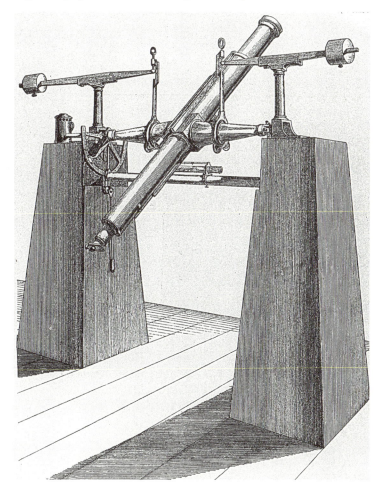

Reichenbach's transit instrument, 1810. Like all transits, the telescope is mounted on balanced cannon-type trunnions between two massive stone blocks. When properly adjusted, the instrument can move only in a meridian great-circle, with no left- or right-hand motion. An observer would make accurate timings with a regulator clock as stars came to the meridian cross-wires to measure their right ascensions.

In 1804 Reichenbach became a partner in a Munich mathematical instrument-making firm, using a technology modelled on that of the finest English manufacturers, and began to produce instruments of high quality and facility of design. His instruments enjoyed a double excellence, indeed, following his collaboration with his fellow Bavarian, Joseph Fraunhofer. Reichenbach's engineering and precision scale divisions, combined with Fraunhofer's optics, represented a formidable combination. It is hardly surprising that the resulting instruments won international acclaim and virtually founded from scratch a high-quality German astronomical instrument-making tradition by 1826, the year in which both men died.

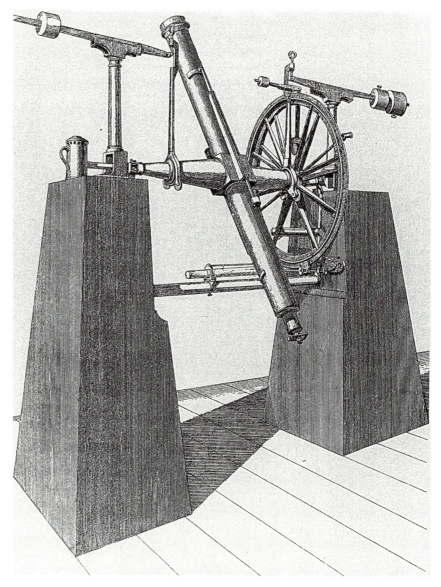

Reichenbach's meridian circle, 1819. Like the transit, the astronomical circle operated only in the meridian, but the addition of an accurate graduated circle to the right-hand axis made it possible to measure the vertical 'declinations' of stars, as well as the right ascensions, in a single observation. For over a century beforehand, right ascension and declination angles had been observed on different instruments, but Reichenbach's engineering innovations succeeded in offsetting stresses within the structure to enable both to be made without distortions occurring.

At the heart of Reichenbach's contribution to instrumentation was his development of the astronomical circle. By 1800, as the success of Ramsden's prototype for the Palermo Observatory came to be acknowledged, the circle had come to displace the quadrant as the astronomers' principal instrument for measuring vertical angles. The full circle of 360 degrees possessed major structural and geometrical advantages over the quadrant. Amongst other things, its shape was more thermally homogeneous than the quarter-circle, so that metal distortions due to temperature were spread evenly throughout its structure and could thereby be made to cancel each other out.

The 360-degree scale meant that one not only had the chance to read off an angle from a single position on the scale, but could place two, four or six micrometer microscopes to read angles at 180-, 90-, and 60-degree spaces. Thus *six* cross-checks could be applied to each observation. These cross-checks enabled the astronomer to conduct exhaustive tests upon the internal geometrical consistency of the scale before a single observation was even made, so that known scale errors could be tabulated in advance, and corrections applied. Theoretically, perfect observations were therefore possible with a circle, as every single source of error could be detected and tabulated in advance so as to obtain flawless declination or vertical-angle values for objects in the sky. Once tested, the circle would be set on a pair of precision trunnions in the meridional plane, to record the declination angles of all objects as they came to the meridian. The best quadrants could measure only down to about 2" arc, whereas the circles of Jesse Ramsden and Edward Troughton, with their multiple cross-checks, could go down to 0.01" arc.

It was Reichenbach's circles which made much of nineteenth-century German astrometry possible, as many of the new and refurbished observatories came to place their orders. Struve had a fine Reichenbach circle at Dorpat, while Bessel at Königsberg did much of his work with one.

The Repsold dynasty

Yet it would be incorrect to assume that Reichenbach enjoyed a field without competitors for the manufacture of precision measuring instruments. Even more significant in the long run was the Repsold dynasty of Hamburg and Bremenhaven, whose three generations of scientist-craftsmen spanned the late eighteenth to the early twentieth centuries. Like Reichenbach and Fraunhofer, the Repsolds were more than just skilled mechanics; they were interested in the problems of physical science in their own right, and were willing to collaborate with professional scientists to produce instruments designed for specific investigations. The father of the firm, Johann

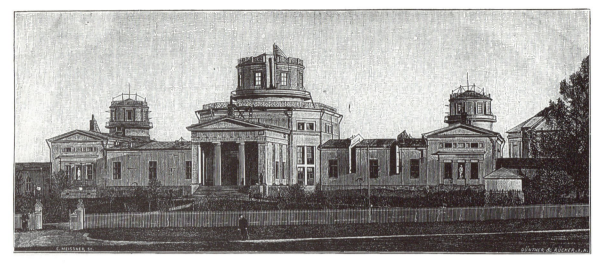

Wilhelm Struve's observatory at Pulkowa, in Russia.

Georg Repsold (1770–1830), corresponded with Gauss about optics, and in 1815 produced a fine meridian circle for his Göttingen observatory. Indeed, Johann Georg received various scientific honours before becoming a hero of Hamburg, following his death in a city fire which he was helping to extinguish.

Johann Georg's son, Adolf (1806–71), carried on the firm, making instruments for Altona, Gotha, and (in 1831) Edinburgh. He received the order for the meridian circle for Struve's new Pulkowa Observatory in 1838, and built another for Bessel at Königsberg in 1839. Adolf Repsold won the contract to supply a heliometer to the Oxford University Observatory in 1842; this became one of the first instruments of that class to be used in England, and it is still preserved in the South Kensington Science Museum.

Fraunhofer's refracting telescopes

The most famous of all German scientist–craftsmen, who revolutionized the making of astronomical instruments, was Joseph Fraunhofer. Although largely self-educated at first, he turned his mundane juvenile apprenticeship as a looking-glass maker to remarkable effect, teaching himself mathematics in his leisure hours, and acquiring a mastery of theoretical optics that was perhaps unparalleled in his day. When this intellectual comprehension of optics is conjoined with his great manual dexterity, one begins to understand his fundamental role in the rise of German astronomy, for he was capable of designing new achromatic object glasses derived from his own experimental investigations, to produce lenses that were perfect in theory and as near-perfect in practice as human hands could make them. Fraunhofer's revolution in designed-lens construction altered the way in which achromatic object glasses had been made, ever since John Dollond invented them in London in 1758. No longer were lenses to be constructed on a semi-empirical basis, as elements were paired and assembled by a refined process of trial and error; their intended functions were now computed before-hand, and their refractive indices and curves were imparted by design.

At the centre of Fraunhofer's achievement was his study of the physical characteristics of optical glass. He quantified refractive and dispersive properties of different glasses to a level of precision never previously attained, so that when the crown and flint glass compo-nents of different achromatic lenses were so designed, there was a much higher degree of control over the intended outcome. Lenses could be designed for specific purposes, and to function best in specific colours or bands of light, depending on the research purposes to which they would be put. It was the pursuit of a mono-chrome light-band, after all, that led Fraunhofer to rediscover and study systematically the spectral lines of sunlight which came to bear his name.

Between 1809 and 1813 in Munich, Fraunhofer collaborated with the Swiss optician Pierre Louis Guinand, to develop Guinand's process of making large-diameter telescope object glasses. This process depended on the production of several optically identical segments of glass, welded together at high temperature to form a single glass blank. The perfected technique made it possible not only to make bigger optical blanks than ever before, but to construct blanks that were free from striae and non-homogeneous regions which gave different refractive indices to different parts of the same lens, with inevitable distortions to the image it produced.

In 1824, Fraunhofer completed a refracting telescope of 9.6 inches aperture for Struve's observatory at Dorpat; this was the largest telescope of the day, and was the admiration of Europe. Fraunhofer pioneered the large aperture refractor which, in the decade before his death in 1826, he supplied to many observatories. It was through one of them, in Berlin, that Galle and D'Arrest first saw Neptune in 1846.

The heliometer

It was also Fraunhofer who pioneered a new class of refracting telescope in the early nineteenth century — the *heliometer*. Basically, the heliometer was a high-quality refracting telescope set upon a clock-driven equatorial mount (see p. 74). Where it differed, however, was the way in which its object glass had been cut in two to produce a pair of semicircular segments. Each segment was set in a semicircular frame, or carriage, at the end of the telescope, so that when a precision micrometer screw was turned, each moved an

ALLAN CHAPMAN

equal distance out of the main optical axis of the telescope. By sliding the segments simultaneously in opposite directions, a split image effect was produced to make the observer see double.

The double image was similar in principle to that produced in a modern photographic range-finder. If the observer wished to measure the angular separation between two stars in the heliometer field of view, he simply adjusted the split images until they appeared to coincide to form a single star. The resulting displacement between the two lens segments, as measured on the micrometer, gave the exact angle between the stars.

In 1837–8, Bessel used a Fraunhofer heliometer at Königsberg to measure the parallax of 61 Cygni, while it was with the Fraunhofer

The principle of the heliometer object glass. Each semicircle was mounted in a precision brass carriage, governed by adjusting screws. When the semicircles were 'opened', a double image was formed in the eyepiece, as in a camera range-finder. But when two stars in a binary pair were made to appear as *one* star, by superimposition, then the displacement of the glasses could be used to compute their angular separation. The lower illustration shows the glasses closed (left), and opened (right).

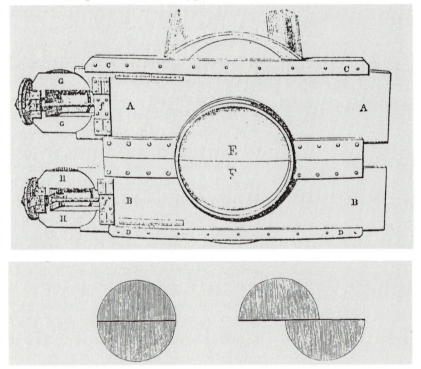

refractor at Dorpat, fitted with an eyepiece micrometer, that Struve also measured the parallax of α Lyra (Vega). It is impossible to underestimate the importance of Fraunhofer's optics to the first successful stellar parallax measures.

After Fraunhofer's death in 1826, Germany was to produce a succession of master-opticians who followed in the path he had first trod. Merz, Petzval, Voigtlander, Zeiss, and many others, produced lenses, both for astronomical use, and for a wide variety of ancillary instruments. It was with a Merz prismatic micrometer fitted to the eyepiece of his own 24-inch reflecting telescope, for instance, that the Liverpool astronomer William Lassell measured the orbits of the moon of Neptune, which he discovered in the late 1840s. After 1850, these German master-opticians further turned their research skills to the development of high-resolution wide-aperture photographic lenses.

The Fraunhofer 'German mount'

In addition to his work on the design of lenses, Fraunhofer applied a great deal of thought and experimentation to developing improved ways of mounting them, so that his long refracting telescopes could be used to maximum effect. Until the time of Fraunhofer, most manufacturers had directed their energies towards the production of better lenses or mirrors, while at the same time being content to permit the fruits of their optical labours to be primitively mounted to the horizontal, or azimuth, plane. It is true that equatorial mounts did exist, and that in mid-eighteenth-century London J. Sisson had developed the 'English mount' for large telescopes; but it was still the azimuth which predominated, being used by William Herschel for all his great reflectors.

Fraunhofer addressed himself to designing a new, and more mechanically efficient, equatorial mount for his instruments. The great advantage of the equatorial mount, as opposed to the azimuth (or altazimuth) mount, lies in its capacity to track objects across the sky around a polar axis, instead of in the horizontal plane. Fraunhofer's solution lay in his elegant *German mount*, whose delicate counterpoises around the polar axis permitted a design which became standard for most big refractors thereafter.

This mount made possible the production of much steadier images than with the azimuth, which needed to be constantly readjusted as the stars rose and set. It also permitted the application of clock-driven mechanisms, which relieved the astronomer of the task of adjusting the telescope once it was fixed upon an object and set in motion.

In addition to his superlative optics, therefore, it might further be suggested that the delicate astrometrical work performed by Bessel, Struve, and Encke's astronomers would not have been possible without the Fraunhofer equatorial mount and its ability to provide steady images of moving stars over many hours of continuous observation.

Conclusion

Germany enjoyed a unique congruence of circumstances which made possible her remarkable astronomical achievements in the nineteenth century. Although Möbius was not an astronomer by vocation, his own mathematical ideas and researches were conducted in a context of intense astronomical creativity, taking place in the discipline in which he held a professorial chair. As Director of the Leipzig Observatory after 1848, he would have been intimately acquainted with the developments discussed above, which he would have passed on in his teaching and mathematical work, and in which his own observatory would have played a part.

Although other countries also made major contributions to astronomy at this time — particularly England, France, and America — Germany's role in its theoretical, observational, and instrumental branches was fundamental. Germany's creativity arose from a confluence of circumstances that derived in part from the country's newly emerging intellectual identity and energy, as well as from an expansion of academic patronage and the appearance of key figures whose enthusiasm and encouragement influenced a whole generation of astronomers. The impact of the 'amateurs' Olbers and Schröter upon the careers of Bessel, Harding, and Gauss was crucial, and from these 'professionals', who combined the talents of both researchers and teachers, there emerged a further and larger generation of talented men. Bessel's observatory at Königsberg became the model of its generation, producing Argelander, amongst others, while Gauss brought lustre to Göttingen, training up Encke, Möbius, and many others. The close relationship which these scientists enjoyed with Fraunhofer, Repsold, and other craftsmen, made possible that highly fruitful union between theory, practice, and the devising of new instruments for the pursuit of specific theoretical goals.

This situation continued virtually up to 1914, as dynasties of scientists and instrument-makers — whether genetic, like the Struve and Repsold dynasties, or pupil-teacher, like Gauss and Encke's discipline — continued to dominate German astronomy.

One might suggest that the only area in which Germany was unequivocally overtaken (and not before the late nineteenth century) was deep-space extra-galactic observational astronomy. In this new branch of the science, the clear desert skies and large industrial endowments of New World millionaires gave America the edge. While German methods of research and instrument design and personnel were evident in American astrophysics, the resurgence of giant reflecting telescopes, under the aegis of George Ellery Hale, showed a clear move away from the favourite German refractor, when the largest optical surfaces were needed.

But Germany's role in the development of astronomy at the time of Möbius was enormous — defining many of the great intellectual issues which came to dominate modern astronomy, creating the instruments and procedures of investigation, and initiating the on-going supply of highly trained professional scientists that we take so much for granted in the modern world.

Acknowledgement

I wish to thank A.V. Simcock, Librarian of the Museum of the History of Science, Oxford University, for his archival and pictorial assistance in the preparation of this chapter.

Further reading

Relatively few books on German astronomy exist in English, but the *Dictionary of scientific biography*, Scribners, New York, has good essay biographies on Möbius, Gauss, Schröter, and Encke.

Nineteenth-century books on the subject include Robert Grant, A *history of physical astronomy*, London, 1852; and John Herschel, *Astronomy*, London 1833. More recent are Henry King, *A history of the telescope*, Griffin, London, 1955; A. Pannekoek, *A history of physical astronomy*, Allen and Unwin, London, 1961; and W. Ley, *Watchers of the skies*, Sidgwick, London, 1963.

Also worth consulting are Joseph Ashbrook, *The astronomical scrapbook*, Cambridge University Press, 1984; and Charles Murray's Halley lecture, *The distances of the stars*, given at Oxford University in 1988 (see *The Observatory*, Vol. 108, December 1988).

Der

barycentrische Calcul

ein neues Hülfsmittel

z u r

analytischen Behandlung der Geometrie

dargestellt

und insbesondere

auf die Bildung neuer Classen von Aufgaben und
die Entwickelung mehrerer Eigenschaften
der Kegelschnitte

angewendet

von

August Ferdinand Möbius

Professor der Astronomie zu Leipzig.

F. Woepcke

Mit vier Kupfertafeln.

Leipzig,
Verlag von Johann Ambrosius Barth.
1827.

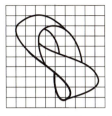

Möbius's geometrical mechanics

JEREMY GRAY

In his day, Möbius was probably best known for his popularizations of astronomy. The discoveries that bear his name today achieved fame and importance only after his death. In between these achievements come the ones that established him among the leading mathematicians of his day: his extensive works on geometry and mechanics.

It might seem that, with the calculus, these were the central areas of mathematics, but few topics in the subject have the curious history that geometry acquired. The Greeks had made geometry seem almost synonymous with mathematics, and for a time the Western revival of the subject adopted this point of view. But it soon merged with another current that may also be traced back to the Greeks, although it has more important and vital sources — algebra. The consequences for geometry were literally overwhelming. During the eighteenth century it almost disappeared, as the language of algebra and the methods of the calculus took over. The geometrical questions the Greeks had asked, and their natural generalizations to new problems, yielded to the new approach, while the old methods seemed increasingly cumbersome and futile. By 1790, the study of geometry had slowed to a murmur.

The history of mechanics was likewise transformed by the advent of the calculus. What can seem obvious, the easy transition between physical problems and their naïve geometrical representations, need not actually be so. Indeed, the leading mathematician of the day, Joseph Louis Lagrange, boasted of having written a treatment of analytical mechanics entirely in the language of the calculus, and without any diagrams at all. That this was doubtless a polemical decision only underlines the point Lagrange wanted to make to his receptive audience: thinking in terms of symbols, formulas, and transformations is the way forward.

All of this changed around 1800, and Möbius's work is very much part of the change. It began in France with the creation of the École Polytechnique and the appointment of Gaspard Monge as its head. Monge was a geometer and a visualizer, who emphasized branches of mathematics such as descriptive geometry, where seeing was essential to doing. His contemporary, Adrien-Marie Legendre, brought Euclidean-style elementary geometry back to the core of the curriculum, driving out what had passed itself off as a Cartesian reliance on intuition about the basics. The result of these decisions, exactly the opposite of British proposals today, was a dramatic

The title page of Möbius's *Der barycentrische Calcul* (1827), in which he introduced his barycentric coordinates based on centres of gravity.

Jean Victor Poncelet
(1788–1867), one of the
creators of modern projective
geometry.

improvement in the quality of French mathematical life. When this improvement was taken to be a principal reason for the success of the French army's temporary conquest of most of Europe, other nations took note.

One of Monge's best students, Jean Victor Poncelet, was taken prisoner by the Russians during Napoleon's retreat from Moscow. He kept up his morale by seeking to extend his teacher's ideas to a new branch of geometry, according to which the key properties of a figure are those that it shares with its shadows. These are its projective properties, and this branch of geometry is now called *projective geometry*. In 1822, some years after his release, he published his *Traité des propriétés projectives des figures*, in Paris. Although controversial, not least because it was obscure, it proved successful, and gave rise to a generation of French projective geometers. At the same time, another French mathematician, Louis Poinsot, took up the idea that there should be a simple geometrical description of how a solid body rotates. For, although Euler had shown how to deal with this problem via the calculus, his treatment was formal and seemed to offer mastery only at the price of insight. Poinsot succeeded, and in a series of publications culminating in his book of 1824, he described the intuitive geometrical ideas that were needed, giving the simple algebra that made a quantified subject, if not of dynamics, then at least of statics.

We shall see shortly how Möbius responded to all of this, but first a word about how he was connected to these French discussions — this was peripherally. The idea that one should be well read in international research journals is a modern, arguably a twentieth-century, one. Möbius moreover was a loner, who preferred to work things out for himself than to read widely. From his arrival in Leipzig in 1816, he seems to have wanted to study mechanics by using geometry and some simple algebra. The building of the observatory there distracted him, and doubtless contributed to his patchy knowledge of the work of Poncelet and Poinsot. Even when the degree of overlap is considerable, it is hard to know whether this is merely in the nature of the subject or reflects some degree of study of the work of others. Since, for that matter, the reception of Möbius's ideas was likewise patchy and hard to disentangle from that of others, it is perhaps more important to observe that it was his combination of geometry and algebra that was to prove potent. This mixture, also found in the work of other German geometers such as Plücker and Hesse, was to prove more vigorous than the purely geometrical approach advocated by Poncelet. By the late nineteenth century, geometry was once again a lively branch of mathematics, and many of its leading practitioners were German, happy to acknowledge the influence of Möbius on the rebirth of their tradition.

Möbius's barycentric calculus

In 1827, Möbius published a little book whose title, *Der barycentrische Calcul*, translates as *The barycentric calculus*. The word *barycentre* means centre of gravity, but the book is entirely geometrical: it deals with lines and conics in the plane and their analogues in space, as well as with certain kinds of transformation of these figures. It is most concerned with, and is best remembered for, introducing a new system of coordinates — *the barycentric coordinates*.

A simple problem

Möbius began with the idea that, if you take a line segment *AB*, two parallel lines *l* and *m* passing through *A* and *B*, respectively, and two coefficients *a* and *b*, then you can find points *A'* on *l* and *B'* on *m* such that

$$a. AA' + b. BB' = 0.$$

How did Möbius solve this simple problem? First, he said, locate the point *P* which divides *AB* in the ratio *b* : *a*, so that *AP/PB* = *b/a*. Then any line through *P* must meet the lines *l* and *m* in points *A'* and *B'* such that the above equation holds; this is a simple consequence of the fact that *PAA'* and *PBB'* are similar triangles.

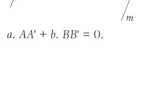

a. AA' + b. BB' = 0.

This seemingly dull problem is typical of his style: Möbius is usually dry this way, but his answer is always clear and simple. Its very dryness should not, however, obscure what is already a novelty in his presentation — the idea of directed (or vectorial) quantities, here symbolized by expressions such as *AA'*, but shortly to be represented by a single letter. However, the idea did not catch on. Nor did it catch on when it was later presented by Grassmann. Its time came only when three-dimensional vectors emerged as three-quarters of Hamilton's quaternions in 1843.

Möbius extended the above problem by showing that you can find further points *A"* on *l* and *B"* on *m* such that

$$a. AA'' + b. BB'' = (a + b). PP'',$$

where *P"* lies on *A"B"*, and *PP"* is parallel to *l* and *m*. To see this, locate *A"*, *B"* and *P"* such that

$$A'A'' = B'B'' = PP''.$$

Then

$$a. A'A'' = a. PP'' \text{ and } b. B'B'' = b. PP''.$$

Adding these to the former result now yields

$$a. AA'' + b. BB'' = (a + b). PP'',$$

as required.

a. AA' + b. BB' = (a + b). PP".

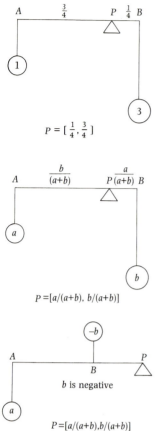

$P = [\frac{1}{4}, \frac{3}{4}]$

$P = [a/(a+b), b/(a+b)]$

b is negative

$P = [a/(a+b), b/(a+b)]$

Three examples of the law of the lever.

Centres of gravity

Then came the analogy that motivated the title of the book. If you imagine a weightless rod with weights of *a* at *A* and *b* at *B*, then, by the law of the lever, the point *P* lies at the centre of gravity of the rod. You can think of the direction of *l* and *m* as the direction of the force of gravity, and of the ratio *b:a* as the *coordinates* of the point *P*.

If the two weights are equal, then the centre of gravity is at the mid-point of the rod. What if they are unequal? Suppose that the weight at *A* is 1, and that the weight at *B* is 3. Then the centre of gravity is nearer to *B* than *A* — in fact, 3/4 of the way along *AB* — because that is the point where a lever with weights of 1 at one end and 3 at the other would balance. In general, if weights *a* and *b* hang from the points *A* and *B*, respectively, then the point of balance divides the segment *AB* at *P*, where *a.AP = b.BP*.

We can also get coordinates for the point *P*, when *P* lies outside the line segment *AB*. For example, to find the centre of gravity when *P* lies beyond *B*, we imagine the lever *ABP* balanced at *P*, with weights at *A* and *B*. One of these weights must be negative — a balloon, if you like — pulling the lever upwards. The same rule continues to apply:

$$a.AP = b.BP.$$

It doesn't matter which of *a* and *b* is negative — it is only their ratio that counts. In this way, points beyond *A* or *B* are given coordinates with a negative entry.

Möbius next showed that his simple argument generalizes to three points. If we fix the weights at *A* and *B* (say, as 1 and 3 again) and hang a non-zero weight from *C*, the centre of gravity *P* will be pulled off the line *AB* and into the triangle *ABC*. Indeed, it will be pulled in the direction of *PC*. The triangle will balance on a knife-edge placed

Weights *a*, *b*, and *c* are attached by strings to *P* and hang from the vertices of the triangle *ABC*. In equilibrium, *P* is at the centre of gravity, or *barycentre* of *a*, *b* and *c*. We say that *P* has *barycentric coordinates* [*a*, *b*, *c*], or [λ*a*, λ*b*, λ*c*] for any λ ≠ 0.

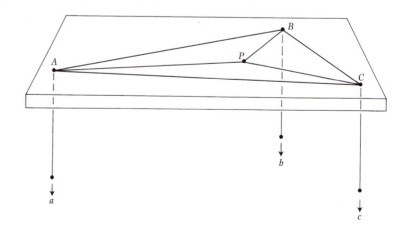

along the line *PC*, as it would along the lines *PA* and *PB*, too. The lines meet at *P*, the centre of gravity or *barycentre*: the whole triangle would balance on a pin placed underneath this point.

The area approach to barycentric coordinates: the barycentric coordinates *a*, *b*, and *c* are proportional to the areas of the triangles *BCP*, *CAP*, and *ABP*.

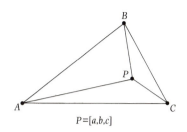

$P = [a,b,c]$

An alternative approach is to start with the triangle *ABC* and observe that any other point *P* generates three triangles — *BCP*, *CAP*, and *ABP*. It is easy enough to show, as Möbius did, that given any three numbers *a*, *b*, *c*, there is a unique position for *P* such that

$$a : b : c = \text{area } BCP : \text{area } CAP : \text{area } ABP.$$

So the ratio of the areas can be taken as the coordinates of the point *P*. In terms of weights and centres of gravity, you take a weightless triangle and hang weights *a*, *b* and *c* from its vertices on fine pieces of string. The centre of gravity (barycentre) of the weights is then at the point with coordinates *a* : *b* : *c* and these coordinates are proportional to the above-mentioned areas.

Möbius showed that any point in the plane is specified by such a set of three numbers — the weights whose centre of gravity lies at the given point — and he proposed that these numbers should be taken as the *barycentric coordinates* of the point. It may seem odd that three numbers should be needed to specify a point in the plane, but actually only the *ratios* of the weights matter: you could replace weights of 2, 3, and 7 grammes, for example, by 2, 3, and 7 tons, or by 20, 30, and 70 tons, and the centre of gravity would still be in the same place.

To distinguish these barycentric coordinates from the more usual Cartesian coordinates, we write them in square brackets; thus the above point would be represented by [2, 3, 7], or by [20, 30, 70], or indeed by [2λ, 3λ, 7λ] for any non-zero number λ. More generally, the point with barycentric coordinates [*a*, *b*, *c*] can equally well be represented by [*a*λ, *b*λ, *c*λ] for any λ ≠ 0; we express this by saying that barycentric coordinates are *homogeneous*. Finally, note that there is one combination of weights that it makes no sense to use — if we place zero weights at each vertex of the triangle, then there is no centre of gravity; thus the coordinates [0, 0, 0] are not allowed.

Barycentric versus Cartesian coordinates

What advantages do barycentric coordinates have over Cartesian ones? To find out, we compare the barycentric and Cartesian coordinates of a point in the plane. To simplify matters, we choose the triangle *ABC* to be isosceles and right-angled, with its vertices at *A* = (1, 0), *B* = (0, 1), and *C* = (0, 0) in the Cartesian plane. We suppose that weights at *A*, *B*, and *C* have their centre of gravity at *P*, with Cartesian coordinates (*p*, *q*). To find the barycentric coordinates of *P*, we find the 'knife-edges' *PA* and *PB* and investigate where they meet *CB* and *CA*, respectively. That will give us points *B'* on *CB* and *A'* on *CA*, and the corresponding centres of gravity or barycentric coordinates.

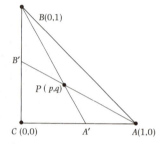

Cartesian coordinates of *P*:
(*p*, *q*).
Barycentric coordinates:
[*p*, *q*, 1 − *p* − *q*].

Consider the line passing through *A* = (1, 0) and *P* = (*p*, *q*). It has the equation

$$y = \frac{q}{p-1}\ (x-1),$$

as you can easily check. So it meets the *y*-axis *CB* at the point *B'*, with coordinates

$$(0,\ \frac{q}{1-p}).$$

This tells us that the length *CB'* is $q/(1-p)$. To find the length *B'B* is then a matter of simple arithmetic — it turns out to be

$$\frac{1-p-q}{1-p}\ .$$

So the ratio *CB'* : *B'B* is $q : 1-p-q$. But the point *B'* lies on *CB*, and so the weight at *A* is 0. Thus, the barycentric coordinates of *B'* are

$$[0, q, 1-p-q].$$

To find the barycentric coordinates of the point *A'* on *CA*, we run through a similiar calculation, switching the roles of x and y. We find that the point A' has barycentric coordinates

$$[p, 0, 1 - p - q].$$

Putting all this together, we find that the barycentric coordinates of the point P with Cartesian coordinates (p, q) are

$$[p, q, 1 - p - q].$$

This must be correct, since the ratio of the first coordinate to the third is the same here as it is for A', putting us on the one knife-edge. Similarly, the ratio of the second coordinate to the third is the same here as it is for B', putting us on the second knife-edge.

Nothing too difficult there — or too interesting! What is interesting, however, is what happens when we try to run this transformation backwards, and determine the Cartesian coordinates of a point whose barycentric coordinates are known.

Let the point have barycentric coordinates $[a, b, c]$. There are two cases to consider.

Case 1: $a + b + c \neq 0$ If $a + b + c = 1$, then we know the answer: to
$$[a, b, c] = [a, b, 1 - a - b]$$

correspond the Cartesian coordinates (a, b). But barycentric coordinates are homogeneous, so we may replace $[a, b, c]$ by

$$\left[\frac{a}{a + b + c}, \frac{b}{a + b + c}, \frac{c}{a + b + c} \right]$$

whose sum is 1. So the barycentric coordinates $[a, b, c]$ correspond to the Cartesian coordinates

$$\left(\frac{a}{a + b + c}, \frac{b}{a + b + c} \right),$$

whenever the denominator $a + b + c$ is non-zero.

Case 2: $a + b + c = 0$. If $a + b + c = 0$, then $ka + kb + kc = 0$ for any number k; thus there is a whole line of points whose barycentric coordinates $[a, b, c]$ satisfy $a + b + c = 0$, but there are no points in the Cartesian plane with these barycentric coordinates. If we call a point defined by its Cartesian coordinates a *Cartesian point*, and one defined by its barycentric coordinates a *barycentric point*, then we deduce the paradoxical result that there are more barycentric points than Cartesian points: the additional points are those for which $a + b + c = 0$.

Möbius spoke of these extra points as points *lying at infinity.* Indeed, if you place such weights at the vertices A, B, and C, then you will find that the corresponding knife edges are parallel.

Projections

We have seen that there seem to be more barycentric points than Cartesian points. This is either a fatal flaw in the whole system — or else it is exciting! Let us see why it is good news.

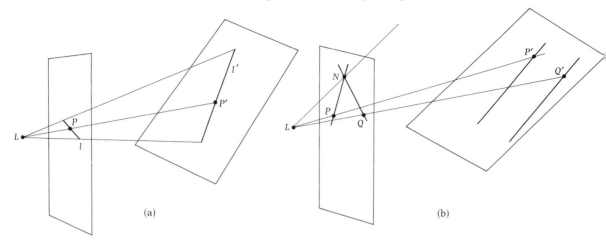

(a)　　　　　　　　　　　　　　　　(b)

(a) Light from a point source L projects the point P and the line l on the first screen to the point P' and the line l' on the second screen.

(b) An example in which the intersecting lines PN and QN on the first screen are projected to parallel lines on the second screen.

Simple shadow projection from a point light source L casts a picture drawn on one translucent screen onto a second screen. The image of a point is a point, and the image of a line is a line. The image of two intersecting lines is usually two intersecting lines, but need not always be so. If the lines meet at the point N, then the image of N is found by following the line LN until it meets the second screen. If, however, the screen is parallel to the line LN, then the lines on the first screen are lines that do not meet — they are, in short, parallel.

The moral of this tale is that there are simple transformations of the plane that 'lose' points. Conversely, if you run the light backwards, you see that the image of two parallel lines can be cast as two intersecting lines: a point seems to have been conjured up from nowhere. There is a traditional form of words that people used to use on such occasions, and which Möbius himself employed. One says that the point of intersection of the parallel lines has been *projected to infinity*.

You might think that, if you follow through these transformations algebraically using Cartesian coordinates, then you will be able to describe where the intersection point has gone. But the argument breaks down. If, however, you use barycentric coordinates, then you can assign coordinates to the image of the point of intersection: there are no missing points.

Let us put the light source L at $(0, -1, 1)$, and investigate the shadow of the points in the xz-plane on the xy-plane. Look at the xz-plane first, and choose a point with Cartesian coordinates (p, q); thus, its x-coordinate is p and its z-coordinate is q. Since its y-coor-

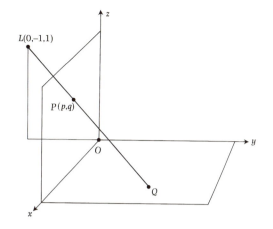

Cartesian coordinates of Q are $(p/(1-q),\ q/(1-q))$; barycentric coordinates are $[p,\ q,\ 1-p-2q]$.

dinate must be 0, we forget about it. So, in the xz-plane, we have a point with Cartesian coordinates $(p,\ q)$, and therefore with barycentric coordinates $[p,\ q,\ 1-p-q]$. It is sent to the point on the line joining it to L that lies in the xy-plane. That point turns out to have Cartesian coordinates

$$\left(\frac{p}{1-q},\ \frac{q}{1-q}\right),$$

for which the corresponding barycentric coordinates are

$$\left[\frac{p}{1-q},\ \frac{q}{1-q},\ 1-\frac{p}{1-q}-\frac{q}{1-q}\right]$$

which simplifies to $[p,\ q,\ 1-p-2q]$. So in barycentric coordinates, the projection sends the point $[p,\ q,\ 1-p-q]$ to the point $[p,\ q,\ 1-p-2q]$.

Now, the points that fail to have an image are exactly those that lie in the horizontal plane through the light source L — namely, those with Cartesian coordinates $(p,\ 1)$ and barycentric coordinates $[p,\ 1-p]$. They are sent to the points $[p,\ 1,\ -1-p]$ — precisely the barycentric points that have no Cartesian equivalents.

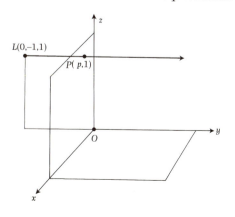

Under horizontal projection from L, the point P has no image.

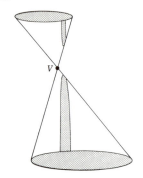

Projection from the vertex V of a cone can send any conic section to any other.

It follows from all this that the study of projective transformations is much easier using barycentric coordinates than using Cartesian coordinates, because you can always see the symbols. It is a balance, really, between the higher price you pay to get started and the ease you have later on.

Duality

There is an excellent reason, well known in Möbius's day, for wanting to study projective transformations. A conic section (an ellipse, hyperbola, or parabola) is a section of a cone. By imagining a light source at the vertex V of the cone, you can see that all sections of the cone are obtainable from each other by projective transformations. With only a little more work, you can see that every conic section is obtainable from any other by a sequence of projective transformations. In particular, every one is the image of a circle. Since circles are easy to study, and projective transformations can be written down explicitly with respect to barycentric coordinates, Möbius was able to reduce the study of conics to the study of circles and projective transformations. This reduction was well known; Möbius's contribution was the simple algebra.

But he went on to do something almost independent and virtually new. As he was finishing writing his book (he tells us in its eventual preface), he heard that in French geometrical work it was proving exciting to associate a line with each point in the plane and a point with each line. The association had to be such that, if you do it twice, you get back to what you started with — that is, if you start with a point you first obtain a line, and then from that line the construction returns you to the original point. Moreover, if three points P, Q, R lie on a line l, then the corresponding lines p, q, r must all have a common point L, and if you start with three concurrent lines p, q, r, then their corresponding points P, Q, R must all be collinear. In a handy piece of jargon, one says that the point L and the corresponding line l are *duals* of each other, and that the correspondence is a *duality*.

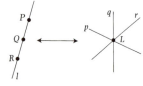

Under duality, collinear points P, Q, and R on a line l correspond to concurrent lines p, q, and r through a point L.

As it happens, there is a neat way of doing this in the plane that goes back to Apollonius, the mathematician who created the fully fledged Greek theory of conics. Start with a conic and choose an exterior point P. Then there are two tangents from the point P to the conic, meeting it at the points S and T, say. The dual of P is the line ST, called the *polar* of the point P. Conversely, the dual of a line l cutting the conic at two points S and T is the point where the tangents at S and T meet. This point P is called the *pole* of the line l. The case where P is an interior point can also be handled — by using a very lovely theorem due to a seventeenth-century mathematician, Philippe de la Hire.

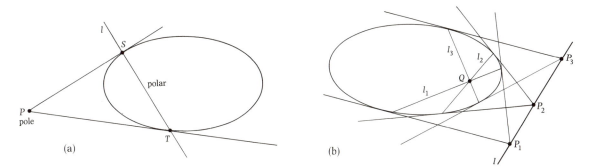

(a) The construction of the pole and polar: the line l is the *polar* of the point P, and the point P is the *pole* of the line l.

(b) De la Hire's theorem: the point Q is the dual of the line l.

Actually, there's something to prove here — namely, that you get the dual of a line l by taking each point on l, obtaining the dual line of that point, and observing that these lines all meet in a common point; this point is the dual of the line l. The fact that you do is the above-mentioned result of de la Hire.

Möbius had a reasonably slick algebraic way of doing all this. Recall that he wrote a line parametrically — the form he used was

$$(1 - \lambda)\frac{a}{p} A - 1\frac{b}{q} B + \lambda\frac{g}{r} C,$$

where A, B, C stand for vectors, which he had written earlier as AA', BB', and CC'. So a point $x\alpha A + y\beta B + z\gamma C$ lies on this line if and only if one can solve the equations

$$x\alpha : y\beta : z\gamma = (1 - \lambda)\frac{\alpha}{p} : \frac{-\beta}{q} : \frac{\lambda\gamma}{r} \ .$$

The letters α, β, γ cancel; they are only there because Möbius liked the generality they convey. Thus, the equations become

$$\frac{x}{y} = \frac{-(1 - \lambda)q}{p} \quad \text{and} \quad \frac{z}{y} = \frac{-q\lambda}{r} ,$$

from which we deduce, after a little algebra, that

$$px + qy + rz = 0.$$

It will help to tidy up his notation in the fashion of later writers. Given fixed values for the Greek letters α, β, γ, Möbius obtained a point for each triple $x : y : z$, and a line for each triple $p : q : r$. In order to produce a duality, Möbius took two planes, with fixed choices for the values of α, β, γ in each of them, asserting the following:

- to each point $u : v : w$ in one plane corresponds the line defined by $u : v : w$ in the other plane;

- to each line $p : q : r$ in one plane corresponds the point defined by $p : q : r$ in the other plane.

This ensures that collinear points go to concurrent lines, and concurrent lines go to collinear points; it also ensures that if you dualize again, you return to your starting point.

Later writers observed that one may think of a line as having coordinates. Indeed, if the line has equation $px + qy + rz = 0$, we let its coordinates be $\{p, q, r\}$, introducing curly brackets to stress the fact that this is a line. Now the duality is as easy as it can be. To the point with coordinates $[a, b, c]$, we associate the line with coordinates $\{a, b, c\}$. Conversely, to the line with coordinates $\{a, b, c\}$, we associate the point with coordinates $[a, b, c]$. This gives us everything we require of a duality, and the algebra could not be easier.

There are problems with this that lie just under the surface, but they can be left there because they do no harm. Note, however, that the coordinates Möbius is using are no longer barycentric coordinates but something a little more sophisticated, called *projective coordinates*, which he explained in his book. They shared several properties with barycentric coordinates: each point of the plane is described by a homogenous triple of numbers; there is a reasonable way to get from Cartesian coordinates to the new coordinates and back; and the equation of a line is of the form

$$ax + by + cz = 0.$$

Mobius's introduction of projective coordinates.

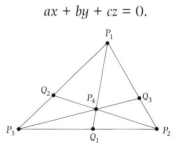

Möbius's introduction of projective coordinates took the following form. He took four arbitrary points, no three of which lie on a line, as a quadrangle of reference, and observed that by joining up the three diagonals one obtains three more points.

By choosing other sets of four non-collinear points from among the points thus obtained, and repeating the construction indefinitely, Möbius obtained what he called a *net* of points and lines (see the box below). By keeping careful track of how sets of four collinear points are generated in the net, Möbius calculated their cross-ratios, which are always rational numbers. As he showed, *a projective transformation preserves the cross-ratio of four collinear points*. He deduced (by an implicit limiting argument) that his net allows any set of four collinear points to be assigned a number (its cross-ratio) that is projectively invariant. So the net plays the same role in projective geometry as would be played in Euclidean geometry by pairs of points a rational distance apart on lines with rational slopes.

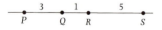

If P, Q, R, and S are four points lying on a line, then their *cross-ratio PQRS* is the number

$$\frac{PR}{QR} \div \frac{PS}{QS};$$

for example, the cross-ratio of the above points is

$$\frac{4}{1} \div \frac{9}{6} = \frac{8}{3}.$$

Möbius nets

Start with any line segment *AG* and extend it to an arbitrary point *H*. Choose another line through *A* and mark on it arbitrary points *E* and *F*. Join *FG* and *EH*, and let them meet at *K*. Join *GE* and *AK*, and let them meet at *C*. Draw the line *FC*, and let it meet *AG* at *B*. Then the cross-ratio *AGBH* is always equal to −1. Let *HC* meet *AE* at *D*. The quadrilateral *ABCD* is the first hole of the Möbius net.

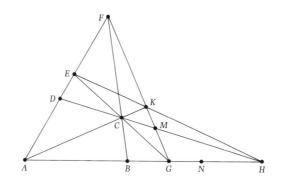

Let *M* be the point where *GKF* meets *HCD*. Using a projective transformation, map the quadrilateral *ABCD* to *BGMC*. Then *G* is mapped to a point *N* such that the cross-ratios *ABGH* and *BGNH* are equal. Repeating this trick along the lines *AG* and *AF*, we obtain as many copies of the hole *ABCD* as we wish. This is the *Möbius net*: it is the projective equivalent of squared paper.

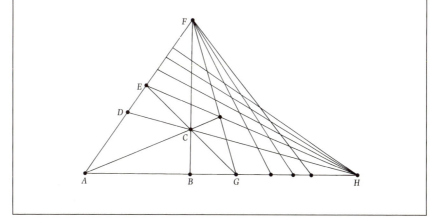

Moreover, coordinates can now be imposed on the projective plane in a way that is invariant under projective transformations. It is enough to join the point to two of the three diagonal points of the quadrangle of reference, and to consider the cross-ratios on two of the opposite sides. These cross-ratios will then serve as the coordinates.

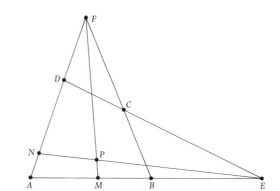

The cross-ratios *AMBE* and *ANDF* serve as coordinates for the point *P* with respect to *ABCD* as quadrangle of reference.

This concludes our account of Möbius's work on geometry in 1827. It was well received in its day, but it was overtaken by the work of Plücker in the 1830s. Subsequently, his later discoveries, to which we presently turn, led mathematicians like Clebsch and Klein back to them, and they acquired a new lease of life, which they continue to enjoy to this day.

Julius Plücker (1801–68).

The Möbius band in projective geometry

Before turning to Möbius's contributions to statics, we digress briefly to consider the appearance of the Möbius band in projective geometry. When an asymptote of a hyperbola is thought of as a projective line, we can regard it as a closed path: we can go out to infinity along it, and come back into the finite part of the plane at the other end. If we do so, we observe that the arm of the hyperbola originally on our right (as in the figure) appears on our left when we return from infinity. The hyperbola is a tangent to the asymptote at infinity, but does not cross it.

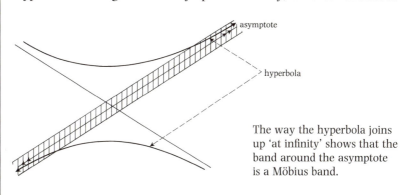

asymptote

hyperbola

The way the hyperbola joins up 'at infinity' shows that the band around the asymptote is a Möbius band.

We now thicken the asymptote, and look at its boundary. As we proceed along the asymptote towards infinity, we consider the boundary curve on our right. From what we have said about the hyperbola, the boundary curve appears on our left when we return from infinity. One circuit of the thickened asymptote yields only half a tour of the boundary curve: the thickened asymptote is a Möbius band! The non-orientable nature of the real projective plane was not to emerge until the early 1870s, with the work of Klein and Schläfli.

Möbius's statics

During the 1830s, Möbius applied himself to the study of *statics*, eventually publishing his book on it in 1837. This, or rather its more vigorous off-shoot *dynamics*, is a topic that had been studied since the time of Newton, and it had become thoroughly assimilated as part of the calculus — specifically, the theory of differential equations. So there was a rich theory devoted to such questions as: *what effect do certain forces have on the behaviour of a rigid body?* The subject matter of statics is also forces, but here the question is: *what forces balance a given set of forces?* It is a subject that was revitalized in 1824 by the French mathematician Poinsot, who saw how to provide a simple, intuitive, geometrical account of the subject that rescued it from the calculus machine.

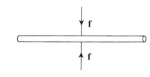

Two equal and opposite forces.

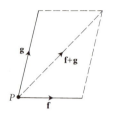

Two forces acting in different directions.

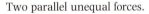

Two parallel equal forces.

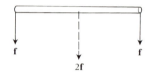

Two parallel unequal forces.

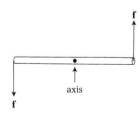

Two equal forces forming a couple.

The picture I want you to have of a force is as a little rocket momentarily squirting out a jet of particles — or, more simply, as a push in a particular direction. In either case, we may want to know where the force is applied. Poinsot directed attention to the question: *given several forces, what is the simplest combination of forces having the same effect?* or *what is the simplest set having exactly the opposite effect?* Such a second set would exactly counter-balance the first, which is why we speak of *statics*, or *statical equilibrium*: nothing moves. For example, if a force is met head-on by an equal and opposite force (one of the same size acting in the opposite direction), then the effect is zero. It follows that a force may be said to act anywhere along its line of action — just think of two equal and opposite pushes down a rod.

The simple case of two forces acting in different directions in the plane is handled as follows. As we have seen, these forces may be supposed to act anywhere along their line of action. Those lines of action meet at a point P, so it is enough to understand that set-up, but that is handled by the *parallelogram law* — namely, if the forces are $\mathbf{f}$ and $\mathbf{g}$, then the combination is given by $\mathbf{f} + \mathbf{g}$.

How about the parallel case? Consider first two equal forces, with the same size and the same direction, applied at different points. If you draw them pointing downwards at the ends of a lever, the answer is clear: they have the same effect as one force, whose size is the sum of their sizes, acting in the same direction through the mid-point. Next, consider unequal forces, with different sizes but the same direction, applied at different points. By the law of the lever, this is balanced by a force in the opposite direction, whose size is the sum of their sizes, and whose point of application is at the barycentre.

More interest attaches to the case of two forces, equal in size but opposite in direction, applied to two different points. Plainly there is no equilibrium, since any body subject to these forces would begin to spin. The forces cannot be reduced in their effect to that of a single force. Such a combination was called, by Poinsot, a *couple*. The part of the lever between the two points of action is called the *arm* of the couple, and the line through the midpoint of the arm (and about which the arm would like to rotate) is called the *axis* of the couple.

We may ask whether there are other couples with a different arm on the same lever and having the same effect. Suppose that we are given opposite forces $\mathbf{f}$ acting at the points P and P'. We consider new forces $\mathbf{g}$ acting at Q and Q'. Plainly the upward and downward forces balance; the question is whether the $\mathbf{f}$-couple balances the $\mathbf{g}$-couple. Again, the law of the lever applies, and the couples balance if the product

(size of force) × (perpendicular distance apart)

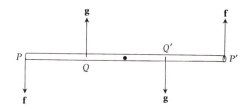

Two couples in equilibrium.

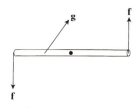

A couple and a force acting in the same plane.

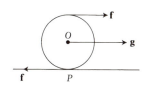

The statics of a rolling wheel.

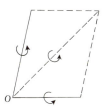

A force acting in space reduces to a force about a fixed point O and a couple.

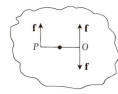

The parallelogram law for couples.

is the same in each case. This quantity measures the size of the couple, and has a technical term, the *moment*. We may rotate the arm of the couple, provided that we keep its moment unchanged — so we can always assume that the couple acts perpendicularly to its arm. As a result, *adding two couples in the same plane reduces to adding forces at the ends of a standard arm: the result is a couple.*

Finally, what can we say about a couple and a force acting in the same plane? This reduces to asking about two unequal forces acting in opposite directions at the end of an arm. By rotating the arm, we can reduce this to the case where all the forces are parallel. By replacing the force by two forces at the ends of the lever, we reduce this to the case of unequal forces at opposite ends of a lever, and thus to a couple. A rolling wheel illustrates this reduction: the natural way to see it is as driven by a force acting horizontally through the axis O and a couple rotating it about the axis. But the view from the point P instantaneously in contact with the ground is different: that point is at rest, and the wheel is momentarily rotating about it.

Let us now move up to three dimensions. Suppose that you are given two forces f and g in space, acting along different lines and applied at different points. What happens then? Let the first force f act at the point P. Pick a point O. Then the force f at P is equivalent to the forces f at P, f at O and $-f$ at O. This combination can be thought of as a force f at O and a couple, f at P and $-f$ at O. Similarly, the second force g acting at a point Q is equivalent to a force g at O and a couple, g at Q and $-g$ at O. It follows that any set of forces reduces to a bunch of forces acting through O and a bunch of couples. Since we can add the forces, the question reduces to combining the couples.

Since a single force is equivalent to a different force and a second couple with a different arm, we can further reduce this configuration to a force and a bunch of couples whose arms all have their midpoints at O. Take two such couples. As the diagram shows, if you draw a parallelogram whose sides lie along the axes of the couple, and whose sides are equal to the moment of the couples, then the diagonal of the parallelogram is fixed — it is the axis of the resultant couple. In fact, couples also behave like vectors.

Lehrbuch

der

S T A T I K

von

August Ferdinand Möbius,

Professor der Astronomie zu Leipzig, Correspondenten der Königl. Akademie der Wissenschaften
in Berlin und Mitglied der naturforschenden Gesellschaft in Leipzig

Erster Theil.
Mit zwei Kupfertafeln.

Leipzig
bei Georg Joachim Göschen
1837.

The title page of Möbius's two-volume *Lehrbuch der Statik.*

We conclude, with Poinsot, that *any combination of forces reduces to a force through an arbitrary point O and a couple about an axis through O.* If the force lies along the axis of the couple, that is the end of the story. If it does not, we resolve it into components along its axis, and perpendicular to it. The component perpendicular to the axis lies in the plane in which the arm of the couple would like to move, so there we have a force and a couple: this is equivalent to a couple. We are left with the component along the axis. We conclude that *any combination of forces acting on a solid body has the same effect at an instant as a couple and a force acting along the axis of the couple.* This may be compared with the theorem that any instantaneous motion is a screw — that is, a rotation around an axis and a simultaneous translation along it.

Statics and geometry

What has all this to do with geometry? This is where Möbius's liking for simple algebra came in. We start, as he did, with the two-dimensional situation. Take a force (X, Y) acting at the point $A = (x, y)$. This is represented by **a**, the vector (X, Y) based at (x, y) and going to the point $B = (x + X, y + Y)$. *What is its moment with respect to the origin O?* The moment is found by dropping the perpendicular from O to the line AB of action of the force, meeting it at a point D. There is some force at O to stop it moving, and the picture is of the arm OD about to rotate. The size of the moment is $AB \times OD$, which is twice the area of the triangle OAB. In terms of (x, y) and (X, Y), the area is $\frac{1}{2}(xY - yX)$ (by a piece of standard coordinate geometry), and so the moment is $xY - yX$.

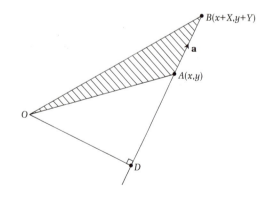

The moment of the force $a = (X,Y)$ acting at the point $A = (x,y)$, with respect to the point O, is twice the area of the triangle OAB.

Möbius showed that if you have forces $(X_1, Y_1), \dots , (X_n, Y_n)$, acting at the points $(x_1, y_1), \dots , (x_n, y_n)$, respectively, then they are in equilibrium provided that

$$X_1 + \dots + X_n = 0, \; Y_1 + \dots + Y_n = 0,$$

and
$$(x_1 Y_1 - y_1 X_1) + \dots + (x_n Y_n - y_n X_n) = 0.$$

This makes sense: there must be no resultant force in the first or second coordinate directions (and hence in any direction), and there must be no resultant moment.

You can interpret these equations as giving the single force that has the same effect at O. Suppose the given forces yield the equations

$$X_1 + \dots + X_n = A, \; Y_1 + \dots + Y_n = B,$$

and
$$(x_1 Y_1 - y_1 X_1) + \dots + (x_n Y_n - y_n X_n) = N.$$

You want a force (X, Y) based at (x, y) which brings the whole lot into equilibrium. Plainly, you must have $X + A = 0$, so $X = -A$, and similarly $Y = -B$. Also, the moment $xY - yX$ must cancel N, so

$$xY - yX = -N, \text{or } xA - yB = N;$$

this tells you the line along which the base point of the force must go. Observe that if you try to balance a couple with a force you will fail, for a couple satisfies $A = B = 0$, and $N \neq 0$. But you can balance a force with a certain moment by another force and a triangle of equal but opposite area.

Let us now move up to three dimensions. The question is: *given a force AB and an axis PQ skew to it, what is the moment of the force about the axis?* Möbius found, by a reasonably simple calculation, that it is six times the volume of the corresponding pyramid. This is a generalization of the previous case. He deduced that a set of forces $(X_1, Y_1, Z_1), \dots, (X_n, Y_n, Z_n)$, acting at points $(x_1, y_1, z_1), \dots, (x_n, y_n, z_n)$ respectively, are in equilibrium provided that

$$X_1 + \dots + X_n = 0, \;\; Y_1 + \dots + Y_n = 0, \;\; Z_1 + \dots + Z_n = 0,$$
$$(y_1 Z_1 - z_1 Y_1) + \dots + (y_n Z_n - z_n Y_n) = 0,$$
$$(z_1 X_1 - x_1 Z_1) + \dots + (z_n X_n - x_n Z_n) = 0,$$
$$(x_1 Y_1 - y_1 X_1) + \dots + (x_n Y_n - y_n X_n) = 0.$$

This also makes sense: the last three quantities are the moments with respect to the x-axis, y-axis, and z-axis, respectively.

Möbius remarked that the above system is in equilibrium if its projection onto each of the three coordinate planes is in equilibrium. He also deduced a result proved earlier by Chasles: *if a system of forces is equivalent to two forces, then these two forces are far from unique, but any suitable pair gives rise to a pyramid of the same volume.* Thus far, Möbius was following closely in the footsteps of his French predecessors, who had earlier used the method of decomposing forces and moments into components and reducing any combination of forces to a force and a couple.

Möbius's contribution

Suppose now that you are given a system of forces in space, once and for all. Pick a point M and look at the lines through it. *How*, Möbius asked, *does the moment of the given system vary with respect to these lines? Along which line is the moment greatest, and along which line is it least?* If the point M lies on the axis of the screw, the answer is clear: the line for which the moment is greatest is the axis of the screw, and the lines perpendicular to that axis have zero moment. With respect to any of these lines, the system of forces does not try to produce a rotation *about* the line, but a rotation *of* the line. If the point M does not lie on the axis, we can re-express the screw in terms of a force through M and a couple with a new axis along the forces through M. That is because the earlier choice of the point O was arbitrary. But the direction of the new axis is different from the old one. Möbius investigated how it changed, and came up with the following lovely description.

The situation is unaltered by moving the force or the couple up and down, and it is rotationally symmetric about the old axis. As you move radially outwards, the position of the new axis gradually tips round as if pushed by the couple, and if you go far enough away, then the new axis is arbitrarily close to the horizontal. All points of space on the same cylinder centred on the old axis have their new axes touching the cylinder.

So, at each point of space, the position of the axis there determines the line of greatest moment, and the corresponding plane perpendicular to the axis is made up of lines through the point for which the moment is zero. Möbius had a useful name for that plane: he called it the *null-plane* of the point. He then asked the converse question: *is every plane in space the null-plane of one of its points?* Pick a plane and look at all the lines in that plane. *Are there any for which the moment is zero? If so, do they all have a point in common?*

The answer will be *yes* if we can locate a point in the plane for which the axis of the force system is perpendicular to the plane, since the corresponding null-plane is then the plane we started with. Can we find such a point? The answer is *yes*, and Möbius had some formulas to show this — but let us look at the situation pictorially, even though this is not actually how Möbius did it.

There is a convenient way of thinking of the direction of a line in space: you note the point at which the line meets a very large sphere, sometimes called the *celestial sphere*; after all, Möbius was an astronomer by profession. If you take the given plane and look at all of its perpendiculars, choosing the upward-facing side for definiteness, you find that they all point in the same direction, and they pick out the same point on the celestial sphere.

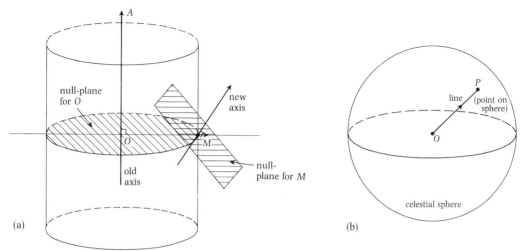

(a)   (b)

(a) Möbius's theorem: every plane is the null-plane of one of its points.

(b) Each directed line in space through O corresponds to a point on the surface of the celestial sphere.

What about the axes of the force system? One axis OA points to the north pole. Since it does not matter what the vertical coordinate of the point is, it is enough to take a horizontal slice and look at all of the axes in that plane. As we saw, any point M at a given radial distance from OA has an axis which points along a line that traces a circle on the celestial sphere; the circle moves steadily down the sphere as the radius OM is increased, until the entire upper hemisphere is covered. So somewhere, one of these axes coincides with the axis of our plane, and we can conclude with Möbius that *every plane is the null-plane of one of its points*; he called this point the *null-point* of the plane.

Now stand back a bit, and forget about the forces: what have you got? *For each point of space, you get a plane through that point, and for each plane in space, you get a point of that plane.* These transitions from point to plane, and from plane to point, are reciprocal. If you start with a point, pass to the corresponding plane, and then pass to the corresponding point — you get back to the point you started with. Likewise, when you go from a plane to a point, and then go to the corresponding plane, you return to the plane you started with. Thus, *by looking at statics, one has obtained a duality* — one that associates to each point of space its null-plane, and to each plane its null-point: the null-point lies in the null-plane, and the null-plane passes through its null-point. What is more remarkable, it is not the type of duality that we discussed earlier.

This discovery, that there is a new and unexpected kind of duality not associated with conics, was one of Möbius's finest discoveries. It so delighted him that he recast it in purely geometrical terms in 1834. In particular, he gave a simple calculation to show that the requirement that each point lies in its dual plane has a simple algebraic interpretation (see the following box).

JEREMY GRAY

Möbius's duality

In Möbius's duality for three-dimensional space, each point corresponds to a plane, and vice versa. Let the point have projective coordinates [**a**]. A plane has equation $\mathbf{b}^T\mathbf{x} = 0$, so we can speak of a plane having coordinates {**b**}. The duality is linear, so the plane corresponding to the point [**a**] must be of the form $\mathbf{b} = \mathbf{A}\mathbf{a}$, for some matrix **A** representing a projective transformation. If the point [**a**] lies on the dual plane, then

$$(\mathbf{A}\mathbf{a})^T\mathbf{a} = 0;$$

that is,

$$\mathbf{a}^T\mathbf{A}^T\mathbf{a} = 0.$$

But if this holds for all points [**a**], then the matrices **A** and $\mathbf{A}^T$ must represent the same projective transformation, as the following little argument shows. Write $\mathbf{a} = \mathbf{u} + \mathbf{v}$; then

$$\mathbf{u}^T\mathbf{A}^T\mathbf{v} + \mathbf{v}^T\mathbf{A}^T\mathbf{u} = 0;$$

so

$$\mathbf{u}^T\mathbf{A}^T\mathbf{v} = -\mathbf{v}^T\mathbf{A}^T\mathbf{u} = -(\mathbf{v}^T\mathbf{A}^T\mathbf{u})^T = -\mathbf{u}^T\mathbf{A}^T\mathbf{v}.$$

Thus $\mathbf{A}^T = -\mathbf{A}$. But we are thinking of these transformations as projective transformations, so **A** is determined only up to a non-zero scalar. So our equation becomes $\mathbf{A}^T = k\mathbf{A}$. What can k be? Since points in projective three-dimensional space are described by homogeneous quadruples, a projective transformation is represented by a 4×4 matrix, so $k\,\mathbf{I}$ is also a 4×4 matrix. So if we take determinants on both sides of the equation $\mathbf{A}^T = k\mathbf{A}$, we obtain the equation

$$\det \mathbf{A}^T = k^4 \det \mathbf{A}.$$

But $\det \mathbf{A}^T = \det \mathbf{A}$, so we can cancel, giving $k^4 = 1$; thus $k = \pm 1$. So **A** must be either a symmetric matrix ($\mathbf{A} = \mathbf{A}^T$), or an anti-symmetric matrix ($\mathbf{A} = -\mathbf{A}^T$)

The symmetric case is the one associated with a conic, or with a quadric surface in higher dimensions. Indeed, the projective equation of a conic can be written $\mathbf{x}^T\mathbf{A}\mathbf{x} = 0$, where **A** is a symmetric matrix. Given a point [**a**], its polar line with respect to this conic is the line with line coordinates {**A**a}, and the equation $\mathbf{a}^T\mathbf{A}^T\mathbf{a} = 0$ says that a point lies on its polar if and only the point lies on the conic and its polar is the tangent to the conic at that point. Exactly analogous statements hold in three or higher dimensions when the equation $\mathbf{x}^T\mathbf{A}\mathbf{x} = 0$ describes a quadric and the duality is between point and plane, or point and hyperplane.

However, the anti-symmetric case is new, and is one of Möbius's finest discoveries. The discovery that in odd-dimensional spaces there is a new kind of duality, not associated with quadrics, is due to Möbius and arose from his study of geometrical mechanics.

The family of lines in space

But Möbius went further: he showed that the above duality can be extended to lines in space. To each line, there is a dual line obtained as follows. Pick each point of the line *l* in turn, and obtain the corresponding null-plane. The intersection of these null-planes is a line, called the *dual line* of *l*. To prove that this intersection is a line is an exercise in duality: because all of the points lie on a line, all of the planes meet in a line.

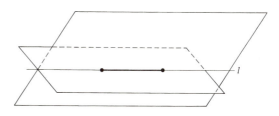

The construction of the dual line of a given line *l*.

Of particular interest are the lines which are self-dual, which Möbius called *double lines*. If a plane contains a double line, then its dual point (as he now called the null-point) lies on the double line. It also follows that all the double lines through a given point lie in a plane, the null-plane of that point.

Attractive though Möbius's duality between lines in space is, there is one drawback: we cannot see the double lines. The best way forward is to try and grasp the space of all lines, and then to pick out the double ones. This step was not taken by Möbius, who was not much attached to geometrical visualization, but it was broached by his successor, the German mathematician Julius Plücker, and became the start of a long story involving Klein and others down the century. The story also embraces optics (the study of light rays under refraction). Moreover, the set of all lines in space is four-dimensional, as we see below. This is agreeably mind-boggling the first time you come across it, because it means that the space we live in can be thought of as not three-dimensional but four-dimensional! This fact greatly excited the educationalist Rudolf Steiner, and remains important in Steinerian thinking to this day.

Suppose that we wanted to draw all the double lines for a given set of forces, fixed once and for all. Through each point of space, there is a plane of them. But each point of space lies on a double line, so our picture is entirely black. On the other hand, not every line in space is a double line — only the lines through the axis of the screw are double lines. This suggests very firmly that we have to study the set of all lines in space.

To see this space, pick a line *l* in space and give it a direction — say, by choosing a unit vector along it. Then, drop the perpendicular from the origin to the line *l*. In this way, you specify a directed line

Mobius's reference to four
dimensions in his *Barycentric
calculus.*

172   Der barycentrische Calcul. Abschnitt II.   §. 141.

einer halben Umdrehung des einen Systems in irgend einer durch die Gerade
gehenden Ebene.

Zur Coincidenz zweier sich gleichen und ähnlichen Systeme im Raume
von drei Dimensionen: *A*, *B*, *C*, *D*, ..., und *A'*, *B'*, *C'*, *D'*, ..., bei denen
aber die Puncte *D*, *E*, ... und *D'*, *E'*, ... auf ungleichnamigen Seiten der
Ebenen *A B C* und *A' B' C'* liegen, würde also, der Analogie nach zu schliessen,
erforderlich sein, dass man das eine System in einem Raume von vier Dimensionen eine halbe Umdrehung machen lassen könnte. Da aber ein solcher
Raum nicht gedacht werden kann, so ist auch die Coincidenz in diesem Falle
unmöglich.

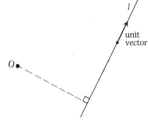

Four numbers are needed to
specify the line l.

uniquely, by specifying the perpendicular from the origin to reach it
and then the direction of the line *l* in space. *How many numbers does
that involve?* Three for the perpendicular, and then two for the unit
vector along the line. (You need two numbers to specify a direction
in space.) However, you must ensure that the unit vector and the
perpendicular are indeed at right angles, so there is an equation to
be satisfied. Five numbers and one equation mean that you need *four*
numbers to specify the line *l*, so the set of all lines in space is four-
dimensional.

To get our first glimpse of this four-dimensional space, we turn the
data around. Take the unit vectors first, and then the perpendiculars. The tips of the unit vectors lie on the unit sphere centred at
the origin; the vectors themselves are radii of the sphere, so the
perpendiculars are all lines tangent to the sphere. In this way, the set
of all lines in space can be seen as the set of tangent lines to the unit
sphere. This object later acquired a name — the *tangent bundle* to the
sphere.

Now go back to the family of double lines. We have seen that
there is a null-plane associated with any given perpendicular. Thus
there is a double line associated with any given direction, for we can
realize it as the intersection of two null-planes. So at each point of
the sphere there is a tangent line representing the given double line.
The family of all double lines realized in this way is what is now
called a *section* of the tangent bundle.

It is worth observing that Möbius's discovery of the above type of
duality could not have been made if he had not been interested in
lines in space. An equation of the form $\mathbf{x}^T\mathbf{A}\mathbf{x} = 0$, where $\mathbf{A}$ is an
anti-symmetric matrix, does not pick out a set of points in space:
every point $\mathbf{x}$ satisfies this equation. The equation does, however,
pick out a set of *lines* in space, the lines which are self-dual, and that
is how it carries a geometrical meaning. For that matter, Möbius's
discovery could not have been made if he had confined himself to the
planar case: there is no such duality there, for simple algebraic

reasons. But Möbius had been led naturally to study three-dimensional space, not just because it is there, but because it is the only interesting space to study if you are interested in statics.

Postscript

These contributions of Möbius invite a few generalizations by way of a conclusion. The attractive and (in their day) important subjects of coordinate geometry and statics were immediately transformed into algebra, the technical means with which Möbius felt happiest. In the geometrical case, particularly, this meant that he did not have to indulge in loose talk about points at infinity, and also that he did not have to base new geometrical researches on the slippery intuitive concepts of weight and force. But his algebra was not merely coordinate crunching. What was clear and new was his emphasis on the linear aspect of the theory — in particular, the realization that something is zero if its coordinates (with respect to some arbitrary choice of axes) are zero. This may seem obvious today, but it is a real question for every physical quantity, and it was a novel mathematical idea at the time. Finally, and most productively, Möbius's working habits demonstrate most clearly how marvellous discoveries can be made by patiently building on the simplest cases and always working fully through the special (and seemingly least interesting) examples. We can all wait for genius to strike, but patient work also brings its rewards.

Further reading

There are few accounts of Möbius's mathematical work on statics and geometry, but he is discussed in the standard history of vector methods: M.J. Crowe, *A history of vector analysis*, Dover, New York, 1985, 2nd edn.

His French precursor Poinsot is discussed in passing in the packed and informative three-volume work by Ivor Grattan-Guinness, *Convolutions in French mathematics 1800–1840*, Birkhäuser, Boston, MA, 1990.

Möbius is better served by German authors. Particularly recommended are E. Scholz, *Symmetrie, Gruppe, Dualität*, Science Networks, Birkhäuser, Boston, MA, 1987; and R. Ziegler, *Die Geschichte der geometrischen Mechanik im 19. Jahrhundert*, Steiner Verlag, Stuttgart, 1985.

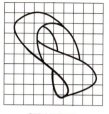

The development of topology

NORMAN BIGGS

In this chapter I shall discuss some of Möbius's contributions to topology, and try to put them in the context of the general history of topological ideas in the nineteenth century. Möbius probably did not think of himself as a topologist, because at that time there was no general subject called *topology*; nevertheless, his ideas have had a profound influence on the development of the subject. The fascinating properties of the Möbius band are described earlier in the book; apart from that, I assume only a very vague knowledge of what topology is about.

Euler's formula

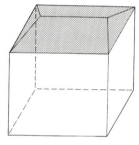

The story begins before the nineteenth century, with the famous mathematician Leonhard Euler (1707–83). Euler made a great number of important mathematical discoveries. One of the simplest, and also one of the most important, was concerned with the vertices, edges, and faces of solid bodies. Since I am not a professional historian, I am allowed to indulge in harmless conjecture, and I suggest that Euler lived in a house that looked a bit like the one shown on the left. It has a certain number *v* of vertices (there are 10 of them), a certain number *e* of edges (there are 17), and a certain number *f* of faces (there are 9). Note that

$$v - e + f = 10 - 17 + 9 = 2$$

in this case. Euler observed that the formula

$$v - e + f = 2$$

is valid for a wide range of objects — pyramids, prisms, crystals of various kinds, and so on — and he claimed that it holds for all solid bodies.

The ramifications of his claim form the main theme of this article, because when people tried to understand in what sense it might be generally true, basic topological ideas arose quite naturally. Euler had a fairly good idea of how he might prove his formula, in some measure of generality, but he did not give what we would now regard as a watertight proof. One of the main reasons for his difficulty was that the basic vocabulary was not developed at that time, and so his claim could not be formulated in an unambiguous way.

The worldwide recognition of the Möbius band is illustrated by its choice as the image on this stamp issued to commemorate the sixth Brazilian Mathematics Congress in Rio de Janeiro in 1967.

In a letter to Christian Goldbach, November 1750, Euler observed that in every solid enclosed by plane faces the sum of the number of faces and the number of solid angles exceeds by 2 the number of edges.

Dieses ist klar, weil keine hedra aus weniger als drey Seiten, und kein angulus solidus aus weniger als drey angulis planis bestehen kann. Folgende Proposition aber kann ich nicht recht rigorose demonstriren:

6. In omni solido hedris planis incluso aggregatum ex numero hedrarum et numero angulorum solidorum binario superat numerum acierum, seu est $H + S = A + 2$, seu $H + S = \frac{1}{2}L + 2 = \frac{1}{2}P + 2$.

Let us move on to the nineteenth century, and a less well-known mathematician who, like Euler, was of Swiss extraction. Simon-Antoine-Jean Lhuilier (1750–1840) worked on this topic for many years, and his most important work was published in 1813. He noticed that certain families of solid bodies do not satisfy Euler's formula, and he set out to give a comprehensive account of these 'exceptions' to the formula; in other words, he attempted to classify those cases where the formula is wrong.

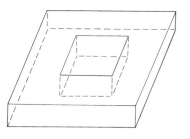

For the sake of variety, let us suppose that Lhuilier lived in a house that looked like the one above; specifically, it has a big courtyard in the middle of it. When we count up the numbers of vertices, edges, and faces, we find that Lhuilier's house has 16 vertices, 32 edges and 16 faces, and thus

$$v - e + f = 0.$$

The reason for this is not hard to see: it is the courtyard in the middle of the house which is causing the problem. The existence of what is essentially a large hole through the solid makes Lhuilier's house in some way different from Euler's house.

Lhuilier recognized the need for a method of classifying solids, in a way which takes account of such problems and makes the significance of the *Euler number* $v - e + f$ clear. Obviously there is some difference between the house with the courtyard and the house without the courtyard, but how should it be described? In fact, there are two questions here. One of them is to explain precisely what is meant by saying that a solid is 'the same' as Euler's house, and the

An East German stamp commemorating Euler's formula.

other is the problem of characterizing the 'difference' between Euler's house and Lhuilier's house in mathematical terms. The first of these problems led to *analytic topology*, while the idea of trying to characterize the differences between solids led to *algebraic topology*. But, back in 1813, there was no topology of any kind.

Despite the lack of a theoretical background, Lhuilier was able to take his ideas a little further. Consider a solid like those discussed above, with vertices, edges, and faces, but suppose that there are g

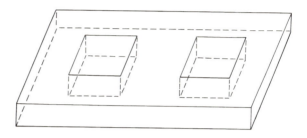

holes running through it, instead of just one hole such as the courtyard of Lhuilier's house. Lhuilier worked out that, in this case, the Euler number is

$$v - e + f = 2 - 2g,$$

whatever the shape of the solid. Nowadays it is said that this is a result about 'topological invariance', but this idea had not been formulated in Lhuilier's time. As always, we must beware of investing historical developments with mathematical hindsight, because the way we think about mathematical ideas now is often entirely different from the way in which the people who discovered the ideas thought about them. Frequently they were studying an entirely different problem from the one which we now regard as the context of their results.

In the early nineteenth century, the obvious questions were rather vague and geometrical. What constitutes a hole in a solid? If there are several of them, how should they be counted? The hole in Lhuilier's house can be thought of as a tunnel, bored out through the middle. If there are several separate tunnels they can be counted quite easily, but things can become more complicated. Somewhere in the interior, there might be a tunnel that links up with another one. More generally, there might be a network of tunnels, rather as if the mice had been at the solid. How should the *holeyness* of such a solid be defined? Why is the value of the Euler number $v - e + f$ relevant? Is that the way to define 'holeyness'?

Möbius and one-sidedness

This is the point to bring in Möbius because, as we know, he spotted another difficulty. All the solids discussed so far, possibly with tunnels running through them, have good old-fashioned two-sided surfaces, they have an inside and an outside. But we know that there are other kinds of surface, because Möbius discovered a one-sided surface, the non-orientable surface we now know as the *Möbius band*. In fact, Möbius did rather more than just give a description of the band. His major contribution was to explain how to describe one-sidedness in a way which is independent of intuitive notions. Indeed, his idea was so fundamental that mathematicians still use it as a definition of *non-orientability*.

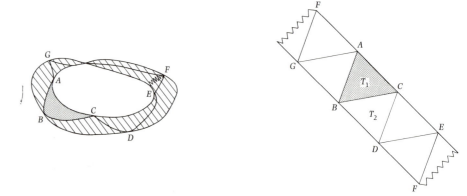

One of Möbius's techniques was to think of a surface as being constructed by glueing together flat polygonal pieces. For example, the Möbius band can be built up from a number of triangles, and when the band is untwisted we can see these triangles quite clearly. The surface of a solid, such as Euler's house, can be thought of in exactly the same way, by adding extra edges where necessary. So the Euler number $v - e + f$ can be defined in rather more general situations. Lhuilier had already investigated similar constructions; in particular, he had looked at solids whose surfaces are not closed, in the sense that (like the Möbius band) they have one or more boundary curves. But Möbius went further, because he was able to explain the phenomenon of one-sidedness in terms of the way in which the pieces fit together.

Let us suppose that we are trying to define the difference between a clockwise rotation and an anti-clockwise rotation. In the above figure, we might pick one triangle T_1 and define the cyclic order ABC of its vertices to be the anti-clockwise one. In the neighbouring triangle T_2, we must clearly choose the compatible rotation, which is *CBD* in this case. The crucial point about compatibility is that, when two triangles share an edge, that edge must be oriented in opposite senses with respect to the rotations of the triangles. Möbius noticed that the triangles forming the band cannot all be given compatible rotations. If we extend the rotations compatibly as far as we can in the untwisted band, then the compatibility condition is violated when we come to construct the band by glueing the ends together. On the other hand, the faces of an 'ordinary' surface, like that of Euler's house, *can* be given a set of compatible rotations. So this is a good way of describing precisely what we mean by saying that the surface of the Möbius band is non-orientable.

Listing's *Census*

That was the great contribution of Möbius to the study of one-sidedness. But not all the credit for the discovery of the Möbius band should go to Möbius. We must also consider the work of a less well-known nineteenth-century mathematician, Johann Benedict Listing (1808–82).

J. B. Listing (1808–1882).

In some ways Listing deserves to be regarded as the founder of topology, not least because in 1847 he wrote a book entitled *Vorstudien zur Topologie*. Indeed, he had already coined the word and used it in correspondence about 10 years earlier. However, there is considerable evidence that the source of Listing's topological ideas was one of the greatest mathematicians of all time, Carl Friedrich Gauss (1777–1855). Listing first studied with Gauss in 1829, and remained in contact with him until Gauss died in 1855. For many years they both worked in Göttingen. In his writings, Listing freely acknowledges that he was trying to develop the topological ideas of Gauss, who himself never published anything on the subject, although there are some relevant comments in his unpublished papers. Gauss's work on the curvature of surfaces and other topics in differential geometry posed many questions which could not be properly formulated without the parallel development of a topological framework. In all likelihood, this was why he encouraged Listing to begin research in what we now call 'topology'.

Let us see what Listing achieved. His first book, the *Vorstudien zur Topologie*, contains interesting (but rather elementary) material, and is not so important from our viewpoint. But another book of his is

The first occurrence in print of the word 'topology' appeared in Listing's *Vorstudien zur Topologie* (1847).

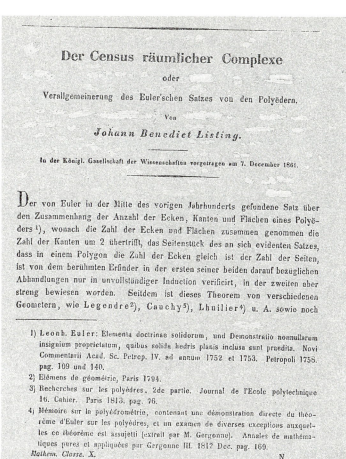

The title page of Listing's
Census.

very relevant: it is *Der Census räumlicher Complexe oder
Verallgemeinerung des Euler'schen Satzes von den Polyëdern* (The census of
spatial complexes or generalizations of Euler's theorem on polyhedra), which was published in 1861. The first noteworthy thing
about it is that it contains a description of the Möbius band. Möbius
himself did not publish the idea until 1865, so there is clearly a
question of priority here. In fact, it appears that both Möbius and
Listing thought about the Möbius band in 1858, because it is
mentioned in unpublished papers of both of them dated in that year;
unfortunately for Möbius, Listing's unpublished note predates
Möbius's by a few months. Both of them describe the construction in
very similar terms. Is this another one of these instances which sometimes happen in scientific discovery, where an idea whose time is ripe
appears independently in different places but at the same time? That
is certainly a possibility. Or was there a common reason for the fact

§. 11. Von der verschiedenartigen Form der zweierlei Zonen-flächen kann man sich eine sehr anschauliche Vorstellung mittelst eines Papierstreifens verschaffen, welcher die Form eines Rechtecks hat. Sind A, B, B', A' (vergl. Fig. 1) die vier Ecken desselben in ihrer Aufeinanderfolge, und wird er hier-auf gebogen, so dass die Kante $A'B'$ sich stets parallel bleibt, bis sie zuletzt mit AB zusammenfällt, so erhält der Streifen die Form einer Cylinderfläche, also einer zweiseitigen Zone, welche die zwei nunmehr kreisförmigen Kanten AA' und BB' des anfänglichen Rechtecks zu ihren zwei Grenzlinien hat. — Man kann aber auch, dafern das eine Paar paralleler Kanten AA' und BB' gegen das andere AB und $A'B'$ hinreichend gross ist, A' mit B, und B' mit A zur Coïncidenz bringen, indem man zuvor, das eine Ende AB des Streifens festhaltend, das andere Ende $A'B'$ um die Längenaxe des

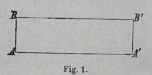

Fig. 1.

The description of the Möbius band as it appears in *Ueber die Bestimmung des Inhaltes eines Poyëders*, 1865.

that both Möbius and Listing described the Möbius band at around the same time? If the latter, then the likelihood is that the common reason was connected with the work of Gauss who, as we know, had been very interested in this kind of topic. Gauss had died in 1855, so the idea cannot have been communicated by him directly, but the possibility of some link with his work remains.

I doubt that the question will ever be completely resolved. Of course, Möbius has several other things named after him, whereas Listing does not, and maybe it would have been fairer if Listing had been credited with the band. However, we shall continue to conform to accepted terminology by calling it the *Möbius band*.

Listing's *Census* addressed itself to the kind of question which arises when we study solids with holes in them and pieces cut out of their surfaces. He was particularly interested in the effect of such operations on the Euler number $v - e + f$. He invented words like *periphraxis* and *cyclosis*, which look like rather serious diseases; in fact, they are his words for topological properties for which we now use other, equally complicated, names. Periphraxis is concerned with the *components* of the surface, and cyclosis is related to what we now call *simple connectivity*. Using such notions, the *Census* describes the peculiar families of objects that Lhuilier had noticed and tries to systematize them.

As I have already mentioned, there were also stimuli from other branches of mathematics in the middle of the nineteenth century. As well as his work on differential geometry, Gauss did several other

This German stamp featuring Gauss was issued in 1955 to commemorate the centenary of his death.

things in which topological ideas (as we now recognize them) were involved. For example, one of his proofs of the *Fundamental theorem of algebra*, that every polynomial equation has a root in the complex field, has some topological ideas in it. Another famous mathematician, Bernhard Riemann, studied *Riemann surfaces*, as we now call them. He was trying to describe geometrically 'functions' which are not true functions — the so-called *many-valued functions*. In order to provide a sound foundation for this work, he had the idea of extending the domain of the function in some way. Instead of defining a function on the complex plane he extended the plane, using geometrical ideas inspired by considering the multiplicities of roots of equations, so that it became a more complicated surface. The point is that, throughout mathematics, the need for a proper development of ideas about topology was pressing. Later in the century, another great mathematician, Felix Klein, developed Riemann's ideas. Together with his colleague Fricke, he wrote a book on so-called *automorphic functions*, which mathematicians still find extremely useful today because it contains so many important ideas about the topology of surfaces and related topics.

We can now see more clearly how the two main threads of topology, the analytic and algebraic, began to develop. First, we have the idea that the Euler number remains constant for solids which have the same number of holes and pieces as the original one — in other words, for solids which are related by a transformation involving no cutting or tunnelling, and which we now call a *homeomorphism*. If we allow cutting and tunnelling, that kind of operation is not a homeomorphism, and consequently the value of $v - e + f$ is not invariant. These ideas were not common currency at the start of the nineteenth century, and a great deal of work had to be done before it became possible to describe in strict mathematical terms what was going on. The other thread is the problem of defining exactly what is meant by the 'number of holes'. That is a discrete problem, made difficult by the fact that the holes may coalesce and join up in curious ways. So it is necessary to turn to algebra, which allows us to make calculations and produce invariants that we can handle.

The introduction of algebraic ideas

The rest of the chapter is concerned with the second thread, the algebraic side of things. There is an interesting general point here about how mathematics develops. How did it come about that, by the end of the nineteenth century, mathematicians were able to describe the 'holeyness' of solids in a proper mathematical way? The strange thing is that the first person to discover a technique for doing this was not specifically a mathematician. He was a famous

G. R. Kirchhoff (1824–1887).

physicist, Gustav Kirchhoff (1824–87), who wrote two remarkable papers in 1845 and 1847, on the flow of current in electrical networks.

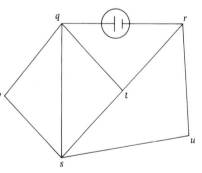

Above is a diagram of a simple electrical network. As a result of the voltage supplied to the network by the battery, some current will flow along the wires, each of which is assumed to have a non-zero resistance. By analysing the network in terms of its cycles, Kirchhoff was able to write down the equations which this flow must satisfy, and to show how they could be solved. It is the second achievement that holds the key to algebraic topology. *Kirchhoff's voltage law* states that the drop in voltage around any cycle of the network is zero;

thus, there is an equation corresponding to each cycle in the network. For example, in the network shown above there are many cycles (such as *pqsp* and *qrtsq*), and for each of them we can write down a corresponding linear equation relating the voltages. The obvious question arises: *how many equations are necessary?* It is clear that one need not write down all the equations, because some of them depend on others. For example, the equation for the cycle *pqtsp* can be obtained by adding the equations for the cycles *pqsp* and *sqts*; consequently, we can say that the cycle *pqtsp* depends on *pqsp* and *sqts*. So the question becomes: *how many of the cycles are independent?*

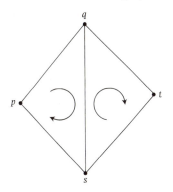

Let us now forget the physical background and consider the network as a geometrical object constructed from points (vertices) and lines (edges). *How many independent cycles are there?* In this case, the answer is manifestly obvious: the four cycles *pqsp*, *qrtq*, *sqts*, and *rustr* are independent, and any other cycle can be formed by putting two or more of them together. But why is the answer *four*? Kirchhoff spotted the fact that it is really 9 − 6 + 1 — that is, it is the number of edges, minus the number of vertices, plus one:

$$\text{number of independent cycles} = e - v + 1.$$

Kirchhoff proved that, in general, this is the correct formula for the number of independent equations. His paper is quite modern in its approach, and he used various constructions which we now think of as standard in graph theory. But he did not have the algebraic techniques that are needed to extend the result to higher dimensions — and that was not part of his programme anyway. However, the basic ideas were latent in Kirchhoff's paper, and it was just those ideas which mathematicians were able to develop in the second half of the nineteenth century, in order to create what we now call 'algebraic topology'.

What Kirchhoff did was to set up the simplest possible framework in which the topological idea of 'holeyness' can be captured in genuine mathematical terms. In fact, his framework is one dimension lower than that needed to describe 'holeyness' in the context of surfaces and Euler's formula. The 'holes' in a network correspond to the cycles, each of which circumscribes an empty region, whereas the 'holes' in a solid are more like tunnels. But the same ideas will work. The way it goes is like this.

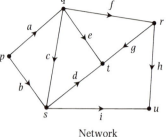

Network

We write down a table, or *matrix*, which describes the network. The rows correspond to the vertices of the network, and the columns correspond to its edges. Because current may flow along a wire in either direction, we assign an arbitrary sense (indicated by an arrow) to each edge; if the sense we have chosen is the 'wrong' one in physical terms, then the value of the current will turn out to be negative. We put

- 1 in row v and column e of the matrix if vertex v is at an end of edge e and the arrow points towards it,

- −1 if vertex v is at an end of edge e and the arrow points away from it,

- 0 if vertex v is not at an end of edge e.

As you can see, each column (edge) has just two non-zero entries, one 1 and one −1. This matrix is known as the *incidence matrix* **D**.

	a	b	c	d	e	f	g	h	i
p	−1	−1	0	0	0	0	0	0	0
q	1	0	−1	0	−1	−1	0	0	0
r	0	0	0	0	0	1	−1	−1	0
s	0	1	1	−1	0	0	0	0	−1
t	0	0	0	1	1	0	1	0	0
u	0	0	0	0	0	0	0	1	1

incidence matrix

One useful property of the matrix **D** is that it enables us to describe cycles algebraically. For example, in the above network, the cycle *pqsp* contains the edges a, c, and b. As we traverse *pqsp*, we agree with the sense of the arrows on the edges a and c, but disagree on the edge b; we represent this cycle *pqsp* by the column vector $\mathbf{x}_{pqsp}$ which has 1 in the first and third positions, −1 in the second position, and 0 elsewhere. Now the rule for operating on a vector with a matrix shows that $\mathbf{Dx}_{pqsp} = \mathbf{0}$, that is

$$
\begin{pmatrix}
-1 & -1 & 0 & 0 & 0 & 0 & 0 & 0 & 0 \\
1 & 0 & -1 & 0 & -1 & -1 & 0 & 0 & 0 \\
0 & 0 & 0 & 0 & 0 & 1 & -1 & -1 & 0 \\
0 & 1 & 1 & -1 & 0 & 0 & 0 & 0 & -1 \\
0 & 0 & 0 & 1 & 1 & 0 & 1 & 0 & 0 \\
0 & 0 & 0 & 0 & 0 & 0 & 0 & 1 & 1
\end{pmatrix}
\begin{pmatrix}
1 \\ -1 \\ 1 \\ 0 \\ 0 \\ 0 \\ 0 \\ 0 \\ 0
\end{pmatrix}
=
\begin{pmatrix}
0 \\ 0 \\ 0 \\ 0 \\ 0 \\ 0
\end{pmatrix}
$$

In fact, every vector $\mathbf{x}$ arising from a cycle in the network satisfies $\mathbf{Dx} = \mathbf{0}$. So the number of independent cycles in the network is equal to the number of independent vectors which satisfy this equation, and nowadays we have a simple algebraic theory which tells us all we need to know about that kind of thing: we talk about *dimensions* of spaces and *kernels* of mappings, and so on. The useful thing is that incidence matrices can be used much more generally to describe how to fit together the pieces of a solid, or indeed an object in any number of dimensions.

Topology into the twentieth century

This development did not happen overnight. The apparatus of vectors, matrices, and what we now call *linear algebra* was not available to Kirchhoff, Listing, and the other mathematicians of the 1840s. That is why Listing's *Census* could not do more than give what we would now regard as a vague and descriptive classification of surfaces. Of course, it is quite clear what he wanted to do — or perhaps we should say, what Gauss wanted to do. What Gauss envisaged became very much the programme of mathematics in the twentieth century. Gauss could not have foreseen the details, but he saw a very long way ahead the outline of what was required.

It is appropriate to conclude by outlining how all this developed into a programme which turned some very vague ideas about the 'holeyness' of solids into an impressive general theory — an algebraic context within which these ideas can be formulated independently of any intuitive notions. There are many famous names associated with this programme. One of them is the Italian mathematician Enrico Betti (1823–92), who introduced numbers, now known as *Betti numbers*, which turn out to be generalizations of the 'Kirchhoff number' $e - v + 1$, the number of independent cycles. But the person who made the greatest advances, in a series of papers

published around 1895, was the French mathematician Henri Poincaré (1854–1912). He took the linear equations and matrices, and put everything into a multi-dimensional context. He explained how one can build up multi-dimensional objects (*complexes*), out of what he called *simplexes*, and he showed how the rules by which they are fitted together can be described by means of matrices. He also showed how the 'holeyness' of complexes can be described algebraically in terms of the properties of these matrices, which are generalizations of the **D**-matrix discussed above. An account of Poincaré's work, which influenced a large number of mathematicians, was published in a volume of the great German *Encyklopädie der mathematischen Wissenschaften* in 1907. The section on topology, written by M. Dehn and P. Heegaard, was a definitive account of the state of the subject at that time.

A little later, but equally influential, was the account given by Oswald Veblen, an American mathematician. At this time, the American mathematical scene was being transformed and was beginning to rival the European one in its general strength. Veblen gave a series of Colloquium Lectures in 1916, which were later published in a book, and in those lectures he gave a modern treatment of Poincaré's theory which stimulated many of the leading American mathematicians to take up the subject.

Throughout the twentieth century the process of generalization has continued. Instead of a matrix or a linear mapping we now talk about a 'homomorphism of modules', and we link these homomorphisms together to form complicated diagrams. The diagrams have properties, such as 'commutativity' and 'exactness', which are in effect expressions of the basic geometrical relationships studied by Lhuilier, Listing, and other mathematicians of the nineteenth century. Unfortunately the background is sometimes forgotten, and the subject becomes merely an exercise in algebraic manipulation. Students who have taken a course in algebraic topology are sometimes unable to see any link between what they have studied and the fascinating properties of objects like the Möbius band. In our earnest endeavours to give students a glimpse of the wonders of modern mathematics, we should not overlook the simple, but equally wonderful, ideas which stimulated its growth.

Further reading

A good introduction to topology, with useful historical background, may be found in M. Fréchet and K. Fan, *Initiation to combinatorial topology*, Prindle, Weber and Schmidt, Boston, MA, 1967. A book written from a different viewpoint, but with much relevant material, is N.L. Biggs, E.K. Lloyd, and R.J. Wilson, *Graph theory 1736–1936*, Clarendon Press, Oxford, 1986; it contains extracts from the works of Euler, Lhuilier, Kirchhoff, and Listing. Euler's formula and its ramifications are central to Imre Lakatos's important philosophical study of mathematical discovery, *Proofs and refutations*, Cambridge University Press, 1976.

A historical account of nineteenth-century topology is given by J.-C. Pont, *La topologie algébrique*, Presses Universitaires de France, Paris, 1974. In particular, this book contains a very detailed discussion and evaluation of the contributions of Lhuilier, Listing and Möbius. General works on the history of mathematics may also be useful — in particular, those by Morris Kline and Howard Eves. The collected works of Euler, Gauss, and Möbius are available in most of the larger mathematical libraries.

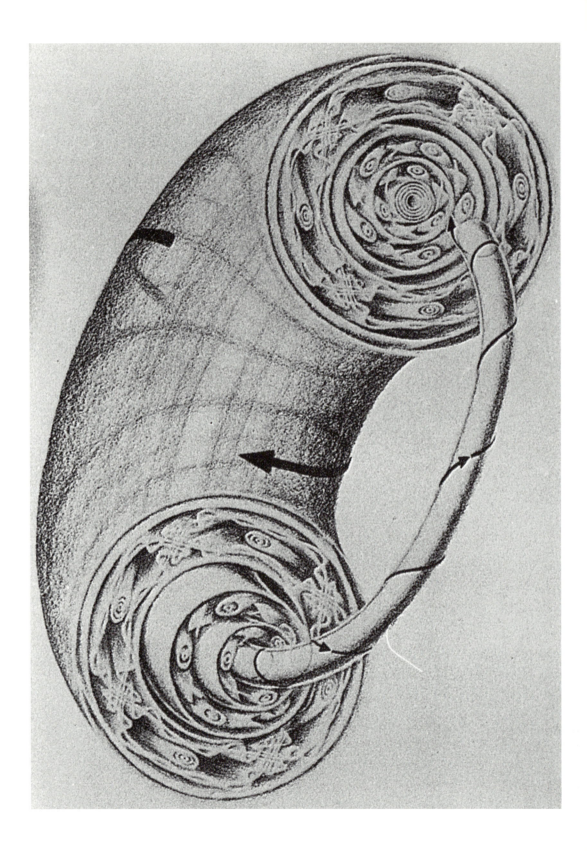

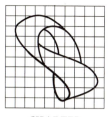

Möbius's modern legacy

IAN STEWART

> While holding the side AB fixed, twist the strip through an angle of 180°
> about its middle line parallel to AB', until A'B' is opposite AB, and then
> bring A'B' into coincidence with A
>
> *One-sided polyhedra*, 1858

Möbius is a household name — at least, it is in mathematical house-
holds — thanks to a topological toy. But August Möbius influenced
mathematics on many levels. Specific ideas — his famous one-sided
surface, his inversion formula, his number-theoretic function, his
transformations of the complex plane, his geometrical nets — bear
his name. But, in addition, and perhaps more importantly, Möbius
was aware of the big ideas, the general principles, the major areas of
research.

What is Möbius's modern legacy? It is a large part of today's mathematical mainstream. The concepts that attracted his attention, and
the methods that he helped to develop, play a central role in modern
mathematics.

Rather than trying to trace a historical path from Möbius to
modern times, however, I'm going to outline some major steps in a
path of ideas. Much of mathematics is communicated by informal
discussions over coffee, seminars, lectures, and other media that do
not produce permanent records. When important mathematical ideas
are 'in the air', other mathematicians get to hear of them by these
informal routes, long before anything appears in a technical journal.
So I'll be using the word 'legacy' in a somewhat loose sense. I mention all this because I don't want to give the impression that Möbius
alone was responsible for the discoveries that I want to talk about.
But he was certainly in there, pushing hard.

I also want to go for the jugular: the *big* ideas that inspired
Möbius — and his contemporaries. Three of Möbius's main interests
were

- topology
- symmetry
- celestial mechanics

To illustrate his influence, I'll concentrate on interactions between
these three areas. It is a measure of the unity of mathematics that
not only *are* there interactions between three such disparate areas,
but that these interactions provide deep and important insights.

The *vague attractor of
Kolmogorov*, an intricate
twentieth-century structure
arising from geometrical
transformations that preserve
area. Möbius was interested
in unusual geometries, and
would surely have approved
of *symplectic geometry* in
which this diagram finds its
natural setting.

Topology

> Two geometrical figures shall be called elementarily related, if ... to each point of one figure there corresponds a point of the other, such that given two infinitely near points of the one, the corresponding points of the other are also infinitely near.
>
> *Theory of elementary relationship*, 1863

The classic potted description of topology is 'rubber sheet geometry', the study of those features of a geometril figure that persist when it is stretched, bent, twisted, compressed, but not torn or cut. The 'rubber sheet' image is not perfect. In particular, you shouldn't think of strong elastic, because topological transformations can stretch distances enormously with no trouble at all; something like infinitely stretchable chewing-gum is closer to the truth. But it's a powerful and vivid image which captures far more than it omits.

Can any features persist under such drastic manipulations? Indeed they can. Möbius's best known discovery, the *Möbius band* (or *strip*), possesses just such a feature, both simple and surprising. It has only one side. This is a topological property, because however much you distort a Möbius band, be it fat or skinny, wiggly or smooth, it *continues* to have only one side. Other properties, such as the band being 2.16 metres long, or having uniform width, or being twisted through so many radians per metre, can be changed by distortions: but not one-sidedness.

Möbius's famous band and its more twisty relatives, from his unpublished writings. The band has two sides if the number of half-twists is even, and one side if the number is odd.

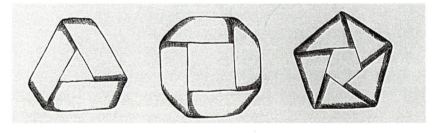

Through the mental eyes of a topologist, a surface formed by such a distortion of a Möbius band isn't another kind of surface with only one side: it's effectively *the same* surface! Topologists train themselves not to recognize non-topological features such as length or width — unless for some reason those features suddenly become relevant to their problem, which sometimes happens. A topologist, so the tale goes, is someone who can't tell a coffee-cup from a doughnut.

This attitude is not as strange as it seems. When you pick up a cup, to drink from it, which of the following alternative descriptions runs through your mind?

(*a*)   the cup I am drinking from is the same cup that was sitting in the saucer;

(*b*)   the cup that I am drinking from is the image, under a transformation that preserves the structure of space–time, of the cup that was sitting in the saucer.

I'm willing to bet that almost all of you answer (*a*) — and for practical daily life, that's the sensible choice. However, when you're doing physics or mechanics, (*b*) is a more precise statement of what you mean by 'the same' when you think (*a*).

Similarly, a topologist who looks at two manifestly different shapes (such as the proverbial coffee-cup and doughnut) and declares them to be 'the same' is tacitly adding 'for the purposes of topology'. You don't find topologists eating coffee-cups or drinking from doughnuts — at least, no more often than non-topologists. A technical analysis of the concept requires a statement closer to (*b*).

• the coffee-cup that I am looking at is the image, under a transformation that preserves the continuity of space, of the standard doughnut-shape, or *torus*, that I find in my textbooks.

By 'preserve the continuity', I mean that points that are close together before the transformation is applied also end up close together. Such a distortion is called a topological transformation; and objects that are the same except for a *topological transformation* are said to be *topologically equivalent* — that is, 'the same for topological purposes'. Möbius's term for this was 'elementarily related'.

To destroy the one-sidedness of the Möbius band, you can take a pair of scissors and cut across it. But points that were initially very close together but on opposite sides of the cut end up a long way apart, so cutting is not a topological transformation.

Just as space–time structure (distance and duration) is crucial to physics and mechanics, so topological structure (continuity) is crucial to topology. What is less obvious, but I hope to convince you is true, is that topology has important consequences for physics and mechanics. The reason is that continuity, as well as distance and duration, is a fundamental property of our universe.

Symmetry

A figure is symmetric (in the broad sense) if it can be placed congruently and similarly in more than one way upon one of its congruent and similar figures.

Symmetry of crystals, 1849

Symmetry is also about transformations — this time from an object to itself. Roughly speaking, a *symmetry* of an object is a way of moving it so that afterwards you can't tell that anything has

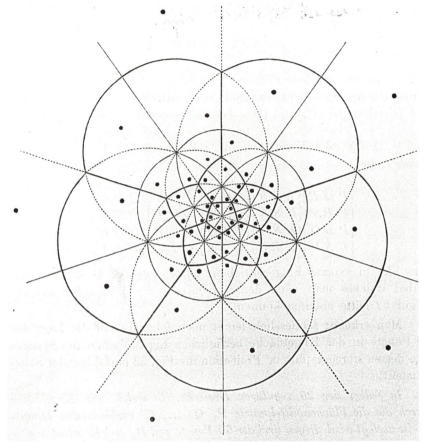

A set of points with the
symmetry of a regular
icosahedron. The points lie
on a sphere, but are drawn
in stereographic projection in
the plane. (From Möbius's
Theory of symmetrical figures.)

happened. For example, a square can be rotated through a right
angle, and still look like a square in the same position. A square has
precisely eight symmetries:

The transformations that are used here are *rigid motions*: all
distances between points of the square remain unchanged. If we
allow other kinds of transformation (think of a rubber square whose
edge is glued down!), we find many more symmetries — in fact,
infinitely many. So symmetries are not arbitrary transformations:
they preserve some kind of structure. What counts as a symmetry
depends on which structure you want to preserve. The mathematician gets to choose which structure; but some structures turn out
to be more useful than others.

Symmetry runs deep in mathematics. One reason is that it offers
enormous simplifications, because symmetrically related objects
behave in similar ways. If you prove something about one corner of a
square — say, that it's a right angle — then the same property of the
other three corners follows 'by symmetry'. That's by no means the
end of the story, however: there's something very fundamental about

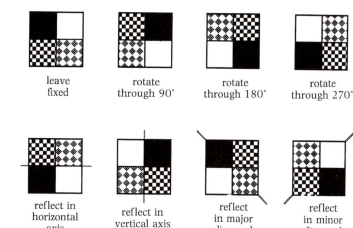

leave
fixed

rotate
through 90°

rotate
through 180°

rotate
through 270°

reflect in
horizontal
axis

reflect in
vertical axis

reflect
in major
diagonal

reflect
in minor
diagonal

The eight symmetries of a square. The effect of each symmetry transformation is indicated by the shading. The first four symmetries are rotations through angles of 0°, 90°, 180°, and 270°; the other four are reflections.

symmetry, and it controls mathematical behaviour in a very strong way. For example, Évariste Galois proved that the quintic equation can't be solved by a formula *because* it has the wrong symmetries. (I bet you never realized that an equation can have any symmetries at all! They are those permutations of the roots — ways of rearranging them — that preserve all valid algebraic relations between them. Once again these are structure-preserving transformations, but in an algebraic setting rather than a geometrical one.)

Celestial mechanics

By the law of universal attraction, the orbit of a planet around the Sun is determined not only by the attractive force that the Sun exerts on the planet, but also by that of the planet on the Sun.

Celestial mechanics, 1843

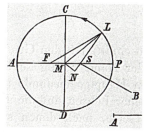

Kepler's first law of planetary motion demonstrated the importance of geometry for celestial mechanics, an idea that continues to play a crucial role to this day. Here we see the elliptical orbit of a planet, reproduced from Möbius's *Celestial mechanics*.

Before the time of Johannes Kepler, the mathematical study of the motions of stars and planets was largely empirical. The situation improved somewhat when Kepler formulated his three basic laws:

First law, a planet orbits the Sun in an ellipse, with the Sun at one focus;

Second law, a planet sweeps out equal areas in equal times;

Third law, the cube of the planet's distance from the Sun is proportional to the square of its orbital period.

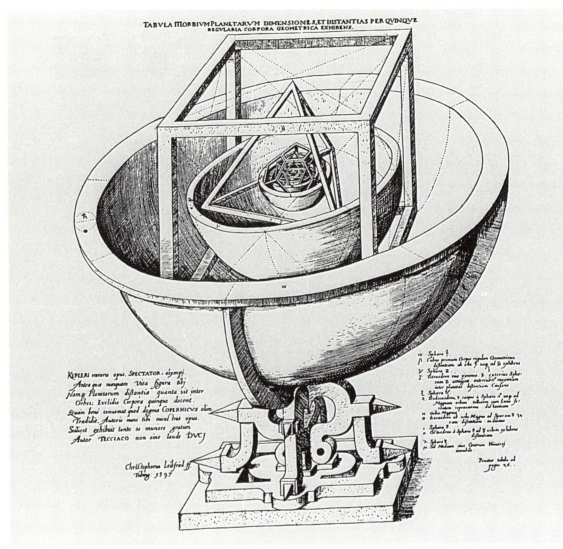

One of Kepler's laws that didn't work out. Attempting to explain why there were six planets (the total known at the time), he suggested that the spacing between them is determined by the five regular polyhedra, in the order octahedron, icosahedron, dodecahedron, tetrahedron, cube.

However, Kepler's writings were far more confused, and confusing, that this list might suggest, and his 'laws' — *now* considered his key insights — are buried amid a host of other speculations and suggestions that have not survived the test of time. For example, he thought that the distances of the planets from the Sun are related to the geometry of the regular solids.

Kepler's laws are not good enough to provide a basis for celestial mechanics. They suffer from two related defects. First, they are restricted to two-body systems: one planet orbiting one sun, or (by analogy) one moon orbiting one planet. They do not obviously extend to more complicated systems, such as star + planet + moon, let alone the Solar System or the Galaxy. Secondly, Kepler's laws are descriptive rather than prescriptive: they tell us whereabouts Mars (say) will be in the sky, but not what physical effects put it there.

Newton removed both defects by proposing two physical principles:

Inverse square law of gravitational attraction, particles attract each other with a force that is proportional to their masses and inversely proportional to the square of the distance between them;

Universality, the inverse square law applies to *every* pair of particles in the universe, and the motion of each particle is determined by adding up all the forces exerted on it by all other bodies.

Newton's principles imply Kepler's three laws, and much more. But they also have a defect of their own — they don't tell us where the gravitational forces come from. Einstein explained that space is bent. Now we're wondering what causes space to bend ...

The precision of Newton's law of gravitation derives from the inverse square law; but its power as a description of *any* gravitating system, not just planet + sun, derives from its universality. For this reason it is often called the law of *universal* gravitation. The point made by the tale of the falling apple is not that Newton wondered why apples fell, and thought of gravity; but that he realized that the *same* force that pulls the apple to Earth also pulls the Moon to Earth, preventing it from drifting away, and thereby keeping it in orbit over our heads.

We might imagine a universe in which the law of attraction was different — say, using the inverse $2^{1/2}$ th power, or the square roots of the masses. In such a universe we could still try to work out whether galaxies exist or whether a planet can have stable rings. But without universality, or something of the kind, we would be unable to study any problems involving three or more bodies.

Geometry of motion

A complete solution of this tangled problem is simply impossible in the current state of analysis..
Celestial mechanics, 1843

To start with, let's look at the interaction between topology and celestial mechanics. The first person to realize just how important such an interaction might be was the great French mathematician Henri Poincaré. Around 1887, he was working on the problem of the stability of the Solar System, for which King Oscar II of Sweden had offered a prize of 2500 crowns. *Will all the planets continue to revolve in approximately their current orbits, or can a planet drift away into the interstellar darkness or crash into the Sun?*

A numerically computed orbit in the three-body problem, for the simplified case in which a body of negligible mass (dust mote) orbits two massive ones (stars). The simple elliptical orbit of the two-body problem is replaced by an apparently structureless tangle. In fact, there exists a hidden geometrical structure, but not of the kind that can be given by a simple formula.

This is a desperately difficult problem. Exactly the feature of Newton's law of universal gravitation that makes it applicable — its universality — also makes the calculations horrendous. Even the motion of three bodies under Newtonian gravitation seemed to lead to insuperable difficulties, as Möbius and many others recognized.

It is said that you can judge the state of advancement of science by the value of n for which the n-body problem is insoluble. In Newtonian mechanics, you cannot solve the three-body problem. In relativity, you cannot solve the two-body problem. In quantum mechanics, you cannot solve the one-body problem; and in quantum field theory, you cannot solve the no-body problem — the vacuum!

If you try to solve King Oscar's problem, then you will quickly realize that every planet, moon, and asteroid in the Solar System is attracted by every other planet, moon, and asteroid — so you can't study separate parts of the system one at a time. Somehow you've got to get to grips with the whole thing at once. Approximate methods do exist that can tell us how the Solar System will move for (say) the next million years or so; but King Oscar was asking about possible nasty events arbitrarily far into the future, so approximations are no use.

In studying this problem (still not fully solved), Poincaré was led to a general investigation of *periodic* phenomena in celestial mechanics — that is, in behaviour that *repeats* after a fixed interval of time. You can see that this might be relevant, because *if* the Solar System were periodic, then no planet could ever wander off or be incinerated — because, by their nature, these are things that cannot be repeated, and certainly not at regular intervals.

Poincaré discovered that he could make a lot of progress if he represented dynamics geometrically. Dynamics is about sets of quantities (positions, velocities) that vary with time: it is not at all clear how to make such things geometrical. Now the Solar System is a big complicated thing, with lots of components. This complexity would divert our attention from the basic simplicity of Poincaré's idea. So instead, let's look at an ecological system with only two variables.

Volterra's paradox

Each new system of axes yields a new viewpoint.

The barycentric calculus, 1827

Around 1925, the Italian biologist Umberto D'Ancona was studying fish populations, and he came across data for the number of fish caught near the port of Fiume, in the Northern Adriatic, for years that included the First World War. He was especially intrigued by the way the percentage of selachians — shark, skate, and ray — varied.

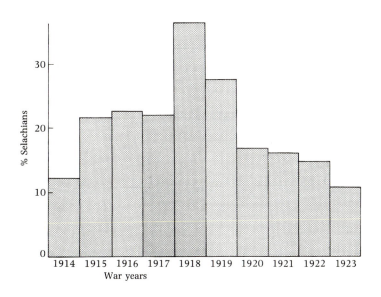

D'Ancona's data show that the percentage of selachians caught in the Northern Adriatic increased substantially during the First World War. Why did a decrease in fishing benefit predators more than their prey?

The percentage of selachians had increased dramatically during the war, and D'Ancona wondered why. Clearly it must be related to the reduced level of fishing in wartime. Selachians are predators — they eat other fish. *Why should a reduction in fishing provide a disproportionate benefit to predators?*

D'Ancona asked the Italian mathematician Vito Volterra for help. Volterra had spent the First World War developing airships as weapons (he was the first to propose using helium instead of inflammable hydrogen), and he now directed his thoughts towards more peaceful pursuits. He devised a mathematical model of the interaction between predators and prey, based on the following arguments.

Suppose at time t there are $x(t)$ prey and $y(t)$ predators. The prey do not compete with each other for food, because there's plenty available. Therefore, in the absence of predators, the rate of growth of the prey is proportional to the number of prey already in existence, so $dx/dt = ax$ for a positive constant a. However, this growth rate is reduced by contact with predators. Assuming random encounters, the number of contacts is proportional to the product xy. So the equation becomes

$$\frac{dx}{dt} = ax - bxy, \text{ for constants } a, b > 0.$$

Predators tend to die off in the absence of food, so with no prey present we expect $dy/dt = -cy$. But contacts with prey lead to an increase in the predator population, and again the number of contacts is proportional to xy; so

$$\frac{dy}{dt} = -cy + dxy, \text{ for constants } c, d > 0.$$

The two displayed equations are known as the *Lotka–Volterra equations* — A.J. Lotka produced them independently in connection with chemical reaction rates. They have a solution that relates x and y:

$$e^{-by} y^a e^{-dx} x^c = K,$$

where K is any constant.

That's the analysis: now we bring in the geometry. We can represent the two variables x and y as the coordinates of a point in the plane. As the time t varies, this point moves, so it traces out a curve. We call this curve the *trajectory* of the initial point (x,y): it's a geometrical picture of simultaneous changes in x and y.

If you plot the family of trajectories given by the above equation, for various values of K, you get a family of closed loops. This means

Graphical representation of population cycles in Volterra's predator–prey model. Each curve shows how the two populations change with time; different starting values yield different curves. Such pictures of dynamical behaviour are called *phase portraits*.

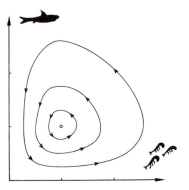

that, no matter what values of x and y you start from, after a certain time you return to the same values; since the curve closes up, it returns to the same point. In other words, the changes in x and y are *periodic*. Thus a dynamic phenomenon (periodic motion) has become a geometrical one (closed loop).

This example contains the germ of the idea which Poincaré made the keystone of his theories of celestial mechanics. But before returning to that topic I'd better show how Volterra's model answers D'Ancona's question about selachians.

From Volterra's equations it can be shown that the average populations of prey and predator, over one cycle, are

$$X = \frac{c}{d}, \quad Y = \frac{a}{b}.$$

By a coincidence — more accurately, because of the rather simple form of the equations — these are the coordinates of the single dot that sits in the middle of all those nested closed loops.

The effect of fishing is to decrease the supply of food fish at a rate ex, and that of selachians at a rate ey, for a constant $e > 0$. The equations change, with a being replaced by $a - e$ and c by $c + e$. The averages become

$$X = \frac{c + e}{d}, \quad Y = \frac{a - e}{b}.$$

That is, increased fishing *increases* the average population of food fish, but *decreases* the population of selachians. Predators and prey are affected in different ways.

This may sound paradoxical, but it illustrates how mathematical models can be superior to verbal ones. *How can increased fishing lead to more food fish?* Easy! The population of predators is decreased, as you'd expect, and that more than compensates for the loss of food fish due to fishing. Conversely, if there is less fishing — as happened during the war — then the number of selachians goes up on average, and the number of food fish goes down. So you catch a greater percentage of selachians.

This result is sometimes called *Volterra's paradox*. It also applies to the use of insecticide to keep down pests. An example occurred in the USA, when the pest ('prey') known as cottony cushion scale insect was introduced from Australia and threatened to wipe out the citrus industry. A natural predator, an Australian ladybird, was introduced, and the population of pests dropped. Subsequently, it was found that the insecticide DDT (now banned because of various unpleasant side effects) killed scale insects, so the citrus industry tried using it to reduce the pest population still further. The population of pests went up!

Volterra could have told them why. DDT also kills ladybirds.

Imaginary spaces

The circle itself can be imaginary.

On conjugate circles, 1858

We now come back to Poincaré. He realized that the same kind of geometrical picture of dynamics can be used in celestial mechanics. Instead of employing just two coordinates x and y for two populations, however, he had to use a large number of coordinates — six per body, in fact. This is because the state of a moving body involves not just its position, but also its velocity. In ordinary three-dimensional space, you need three coordinates to specify position, and three more to specify velocity. So the system Earth + Moon + Sun, for example, involves 18 variables. Thus the simultaneous motion of the Earth, Moon, and Sun can be visualized as the motion of a single representative point in a space of 18 dimensions — assuming that 'visualized' is quite the right word ...

This imaginary space, whose coordinates are the variables that determine the state of the system at any instant, is called *phase space*. The phase space of Volterra's model is a plane, because only two variables are needed to describe the two fish populations. The curves that are traced out by the representative point as the system moves are called *trajectories* (or often, confusingly in the present case, *orbits*). The system of all possible trajectories is the *phase portrait* of the system. In Volterra's model, it is the set of concentric closed curves. The phase portrait has the advantage of capturing, not just one possible motion, but all possible motions, in a single geometrical object. Trajectories resemble the flow-lines of particles in a fluid, so the entire structure is a *flow* of imaginary particles, whose positions represent states of the dynamical system, and the fluid is the set of all possible states — that is, it fills the whole of phase space.

Poincaré's grand scheme is now apparent. Instead of asking analytic questions about solutions of differential equations, we can ask geometrical questions about phase portraits (or flows). You can build up a kind of dictionary to translate from dynamics to geometry. For instance,

- a solution is *steady* if the corresponding trajectory is a single point

- a solution is *periodic* if the corresponding trajectory is a closed loop

- a solution is *quasi-periodic* (composed of several independent periodic motions) if the corresponding trajectory is a spiral that winds around a torus

- a solution is *stable* if all nearby trajectories approach it

Poincaré introduced qualitative geometrical pictures of dynamic behaviour. Concepts such as 'steady', 'periodic', 'quasi-periodic', and 'stable' all have natural geometrical interpretations.

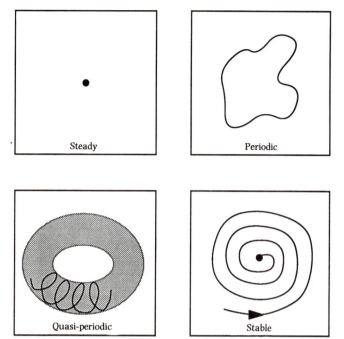

This was Poincaré's vision: a geometrical method for solving difficult problems in differential equations, especially celestial mechanics. He realized that it was going to be a new and different kind of geometry: he called it *analysis situs*, the geometry of position. In 1838, J.B. Listing introduced the name we use today — *topology*. It applies to many different areas of mathematics, not just to differential equations.

There was a snag, of course. A phase space with 18 variables is a complicated beast. Is there really any advantage to be gained by changing from analysis to geometry? Can we comprehend the picture in 18 dimensions? *Can geometry be used to solve problems that analysis can't tackle?*

Morse theory

A method will be developed, by means of which the form of a closed surface can be represented by a simple scheme.

Theory of elementary relationship, 1863

The answer is that it can, and the simplest example lies in an area that greatly interested Möbius — the topology of surfaces. The simplest version concerns not dynamics, but *statics* — the steady states, or equilibria, of a dynamical system. A good example is the double pendulum, in which two rods hang in the plane, with masses attached. For comparison purposes, we also consider a far more complicated system — a double pendulum with various bells and whistles (actually, elastic springs) attached.

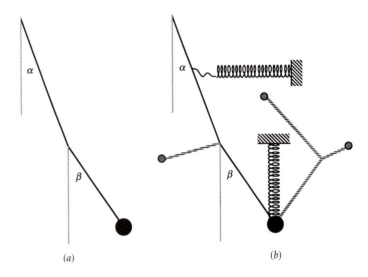

(*a*) The ordinary double pendulum: two rigid rods hinged together in the plane. (*b*) An elaborate cousin, bedecked with springs and elastic. Topological considerations imply that both systems have at least four equilibrium positions.

Because we're doing statics, we don't need to worry about velocities. To make this clear, we talk of *configuration space* instead of *phase space*. For both systems, the configuration space is determined by the two angles α and β. Each individual angle corresponds to a point on a circle, because 0 radians is the same as 2π radians and so the interval of angles between 0 and 2π has to be bent round to join up the ends. Therefore a *pair* of angles corresponds to a point on a torus, the Cartesian product of two circles. Thus the configuration space for a double pendulum, with or without bells and whistles, is an ordinary torus.

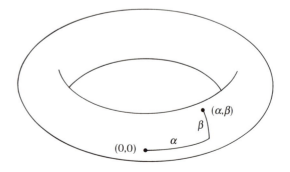

Pairs of angles (α,β) determine points on a torus, just as pairs of real numbers (x,y) determine points in the plane.

The equilibrium positions are determined by the potential energy (including the elastic energy) of the system. If the total potential energy in the configuration (α,β) is $E(\alpha,\beta)$, then the equilibria are those values of (α,β) that make E stationary:

$$0 = \partial E/\partial\alpha = \partial E/\partial\beta.$$

If we draw the graph of $E(\alpha,\beta)$ against (α,β), then these stationary points are places where the graph is 'horizontal'.

For the ordinary double pendulum (a), we can easily find these stationary points by calculus. The energy function is

$$E = -mgl \cos \alpha - Mg \,(l \cos \alpha + k \cos \beta \,),$$

where m and M are masses and l is the length.

The vanishing of the first partial derivatives leads to the equations

$$0 = \sin \alpha = \sin \beta,$$

so that

$$(\alpha,\beta \,) = (0,0),\ (0,\pi),\ (\pi,0),\ \text{or}\ (\pi,\pi).$$

That gives us four equilibria: each pendulum hangs vertically down (0) or up (π), in all four possible combinations.

For the more complicated double pendulum (b), the bells and whistles greatly complicate the formula for $E(\alpha,\beta)$. Indeed, I'm not going to write one down, because it's clear that I could complicate the system sufficiently to make any analytic solution ridiculously messy. Instead, I'm going to show you a simple topological argument which implies that, whatever the bells and whistles may be, there will still be *at least four* equilibria. There could be more, but definitely no fewer. So topology easily gives a partial answer. It doesn't tell us where the equilibria are, nor even their exact number, but it does tell us there must be at least a certain number of equilibria, independently of the system, *because* its configuration space has a particular topological form.

The fundamental theorem of the topology of surfaces tells us that every surface (technically it must be compact, orientable, and without boundary) is topologically equivalent to a sphere with g holes in it. The number g is the *genus* of the surface. A sphere has $g = 0$, an ordinary torus has $g = 1$, a double torus has $g = 2$, and so on. We study systems for which the configuration space is a surface of genus g.

A surface of genus g is one that has g holes. Every two-sided closed surface without boundary is of this type, up to a topological transformation.

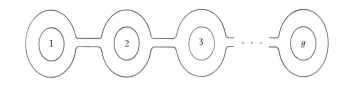

We can visualize the energy function on this surface by distorting the surface in such a way that the values of the energy are represented by height. Any such distortion defines a function (height), and any function can be so represented by choosing the appropriate distortion. The stationary points that we are seeking are the places where the surface has a horizontal tangent plane. Möbius was very interested in the height function for a surface with holes. Our diagram is the same as one that he published in 1863, and the arguments we shall follow are similar to ideas that he developed at that time.

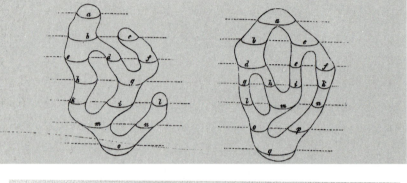

A real-valued function, defined on a surface of genus 1, represented by height. (From Möbius's *Theory of elementary relationship.*)

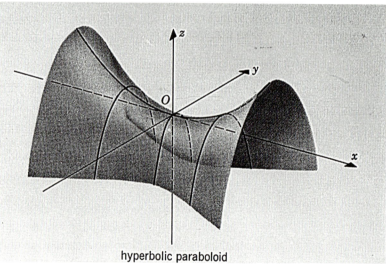

A mathematical saddle has the same properties as the saddle on a horse — starting from the centre and moving from nose to tail means going up, while going from side to side means going down.

hyperbolic paraboloid

There are three basic kinds of stationary point: *maxima*, *minima*, and *saddles*.

Möbius's example has three maxima, one minimum, and four saddles.

In general, suppose there are N_{max} maxima, N_{min} minima, and N_{sad} saddles. I claim that, *whatever the function E,*

$$N_{min} - N_{sad} + N_{max} = 2 - 2g.$$

Proof that $N_{min} - N_{sad} + N_{max} = 2 - 2g$

The idea of the proof is slowly to deform the position of the surface to simplify the calculation of the quantity $N_{min} - N_{sad} + N_{max}$.

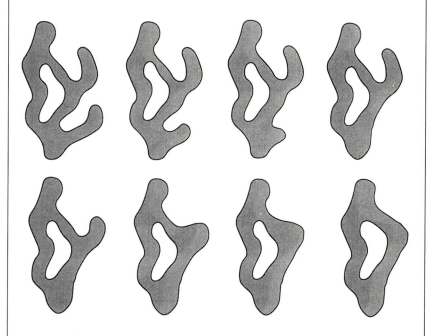

Deforming Möbius's surface into standard position so that it has one maximum, one minimum, and two saddles, by applying a sequence of topological transformations that 'cancel' stationary points in pairs.

We try to move the g holes into 'standard position', one below the other. Most deformations don't change any of the numbers N_{min}, N_{sad}, or N_{max}, but there are two ways in which those numbers do change. In one of them, a maximum and a saddle cancel each other out (or, reversing the deformation, are created from nothing). In the other, a minimum and a saddle behave in the same way.

Cancelling a maximum and a saddle by merging them so that they annihilate each other. To cancel a minimum and a saddle, turn the picture upside down.

Thus those deformations that change the values of N_{min}, N_{sad}, or N_{max} do so in the following way:

increase or decrease N_{max} and N_{sad} by 1,

increase or decrease N_{min} and N_{sad} by 1.

In each case the change in value of N_{max} or N_{min} is exactly cancelled by that of N_{sad}, so the value of $N_{min} - N_{sad} + N_{max}$ doesn't change at all. We say it is a *topological invariant*. But after all the deformations have been carried out, we have a g-holed surface in standard position, for which

$$N_{min} = 1,\ N_{sad} = 2g,\ N_{max} = 1,$$

so

$$N_{min} - N_{sad} + N_{max} = 1 - 2g + 1 = 2 - 2g,$$

as claimed.

For example, in the above surface there is one hole, so $g = 1$, and $2 - 2g = 2 - 2 = 0$. We have $N_{min} = 1$, $N_{sad} = 4$, $N_{max} = 3$; and indeed $1 - 4 + 3 = 0$. Try some other examples of your own, with plenty of holes. It still works!

This is a remarkable formula. It relates dynamics (equilibria — that is, stationary points of E) to topology (the number of holes in configuration space). Using the formula, we can show that N_{min} is at least 1, N_{max} is at least 1, and N_{sad} is at least $2g$. To prove this, note that there must always exist at least one maximum and one minimum, so the first two statements are obvious, and we can set

$$N_{min} = a + 1,\ N_{max} = c + 1,\ N_{sad} = b,\ \text{where } a, b, c > 0.$$

Now we have

$$(a + 1) - b + (c + 1) = 2 - 2g,$$

and so

$$a + c + 2g = b.$$

Therefore $b \geqslant 2g$, which is the third statement.

Whenever the configuration space is an ordinary torus, with $g = 1$, we have at least two saddles, so at least four critical points altogether. This is why I can be certain that the double pendulum with bells and whistles always has at least four equilibria, whatever the bells and whistles may be.

The number $2 - 2g$ is called the *Euler characteristic* of the surface, and it appears in another very similar and very famous formula, which goes back to Leonhard Euler (see Chapter 5). Suppose that we cut the surface up into lots of polygonal regions, so that there are v vertices, e edges, and f regions (faces). Then $v - e + f = 2 - 2g$ as well, however we do the cutting. The connections between these two formulas, and their generalizations to higher dimensions, is called

Morse theory after its inventor Marston Morse. Morse theory is a powerful technique that links dynamics — and more generally any problem involving stationary points of a function — to topology. It is firmly rooted in ideas that Möbius developed about 130 years ago.

Sliced dynamics

It is not difficult to convince oneself of the correctness of these results without carrying out any calculations.

Celestial mechanics, 1843

Poincaré realized that geometry helps with dynamics as well as statics. In particular he devised a topological method for proving that a system of differential equations has a periodic solution. The idea is to take a multi-dimensional sheet (made of rubber, of course, or better still, chewing-gum) that cuts across trajectories. This is now called a *Poincaré section*. Having chosen such a sheet, you start it out at some specific time and let the *entire sheet* flow along the trajectories. You now see why it has to be a rubber sheet, because the flow lines may pull apart, bunch together, or swirl round in complicated ways. The laws of dynamics prevent any tearing or cutting, however: the deformation is always continuous. So we've got a piece of rubber sheet flowing along, deforming as it goes. Suppose that some of it eventually comes round towards the place it started from — where you thoughtfully left a copy of it to remind yourself where it was to begin with. What happens?

Imagine that the travelling sheet is sticky as well as made of rubber — as I said, chewing-gum is a good image — and that the copy is what computer scientists call 'hard copy' — fixed solidly in place. The travelling chewing-gum sheet runs into the hard copy: *SPLAT!* Different parts of it hit at slightly different times; but we don't mind that, we can wait.

At the end of this bizarre ceremony, what have we achieved? We've taken a sheet of mathematical chewing gum, bent it, and

A *Poincaré section* is a surface that cuts across the flow-lines of a dynamical system. The associated Poincaré mapping maps each point on the section to the first point to which it returns after following the flow.

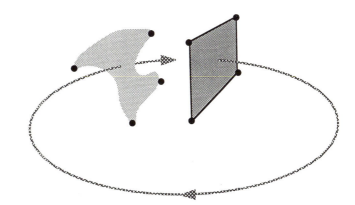

slapped it back on top of itself. In short, we've constructed a topological transformation from the sheet to itself. This is the *Poincaré
mapping* associated with the chosen Poincaré section.

*Can we detect the presence of periodic trajectories using just a Poincaré
section?* We can. Periodic trajectories are closed loops, so they return
to the same point — that is, they meet the section at one point, flow
round for a while, and hit the section again at exactly the same
point. So the Poincaré mapping takes that chosen point, and slaps it
down *exactly on top of itself*; in other words, it is a *fixed point* of the
Poincaré mapping.

This is a great simplification. Instead of looking for periodic solutions of differential equations — which can be hard, and tends to
make you think about complicated formulas — you can look for
fixed points of topological transformations, which is usually a lot
easier. For a start, the section has one dimension less than the phase
space; and transformations are generally easier to think about than
solutions of differential equations.

If all you want is the *existence* of a periodic solution, then sometimes topology can give the answer without any calculations at all.
Suppose that the Poincaré section is a multi-dimensional version of a
disc, with no holes in it; and suppose, in addition, that every point
on this disc eventually slaps into the hard copy. Then the entire disc
is mapped inside itself by the Poincaré mapping. A famous theorem
in topology — *Brouwer's fixed point theorem* — then tells us that
there must be at least one fixed point. That is, in such a situation
there has to be at least one periodic trajectory. So here's an example
of a general topological fact implying something significant about
dynamics.

But the Poincaré section is far more useful than that. It carries a
great deal of information about all of the dynamics near the periodic
trajectory. For instance, if there are several other fixed points, then
we have found several other periodic trajectories. In fact, the
Poincaré mapping tells us how *all* nearby points behave, not at all
times, but at some discrete sequence of times: every time the point
comes back and re-encounters the Poincaré section. We can find the
sequence of points in which its trajectory hits the section, by
repeatedly applying the Poincaré mapping — a process known as
iteration.

For a slightly more exotic example, suppose the Poincaré section
has two points *A*, *B*, such that, under the Poincaré mapping,

A maps to *B*, and *B* maps to *A*.

This is a *period-2 point* of the Poincaré mapping. For instance,
suppose that the Poincaré mapping just rotates everything through
180° about the fixed point; then any choice of *A* is a period-2 point.

In the full dynamics, what do we see? As the point on the original periodic trajectory revolves once, the point *A* revolves nearby, but moves on to *B*. It requires a second revolution of the original trajectory to take it back to *A*. So *A* describes another periodic trajectory, but one that wraps once around, while the original trajectory wraps twice around. We say that is it is in *2:1 resonance* with the original trajectory.

If you draw a line along the middle of a Möbius band, then it wraps only once round while the boundary wraps round twice. So the geometry of the 2:1 resonance resembles a Möbius band. This fact is often important. The Möbius band isn't really a toy — it just looks like one.

There are other kinds of resonance, such as 3:1 or 5:2 or 22:7, which work the same sort of way. Resonances occur a lot in celestial mechanics. For example, Jupiter's satellites *Io* and *Europa* have orbital periods of 1 day, 18 hours, 27 minutes, and 3 days, 14 hours, 13 minutes. The ratio of these two periods is 2.03, so they are very close to 2:1 resonance. Saturn's satellites *Titan* and *Hyperion* have orbital periods of 15 days, 22 hours, 41 minutes, and 21 days, 6 hours, 38 minutes, a ratio of 1.3337. This is very close indeed to 4:3 resonance. Mercury's rotational period is in 2:3 resonance with its orbital period. Those are a few of the details — but let them not obscure the grandeur of the vision: *a 'natural' geometry of dynamics.*

Poincaré could see that many new developments would be needed to make his vision become a useful tool. He even had a good idea what those developments might be — and he knew they would be hard! But his mathematical sense of smell was functioning magnificently: Poincaré's nose had detected a very important idea. He didn't live to see it developed, but in the later years of his life he himself made two important contributions, each taking an almost impossible analytic problem, turning it into something geometrical and natural, and deducing something important and novel. Both ideas have had major consequences for mathematics. One led to *chaos* (a currently fashionable area attracting considerable attention from the popular media), and the other to *symplectic geometry* (which is likely to be even more important, but which you've probably never heard of). Let's take them in turn.

Chaos

But this is not possible.

Textbook of statics, 1837

What kinds of things can Poincaré mappings do? Let's just think about what happens when we iterate a mapping from a surface (such as the plane) to itself. Here are six possibilities (from what, in principle, is an endless list).

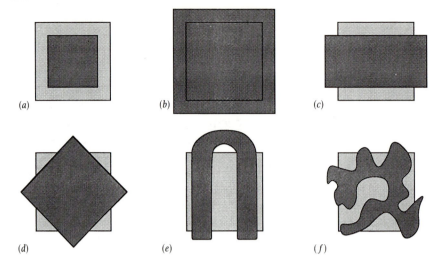

Six mappings of the plane, each having different dynamics:
(*a*) shrink in all directions;
(*b*) expand in all directions;
(*c*) shrink in one direction but expand in another;
(*d*) rotate;
(*e*) twist into a horseshoe shape;
(*f*) combine all previous features arbitrarily.

(a)　is a mapping that shrinks everything. It's clear that if you iterate this, you just shrink things more and more; so all points move in towards the centre, which is the unique fixed point. The associated dynamics consists of a periodic trajectory (corresponding to the fixed point) and everything else contracts down towards that trajectory, while looping around closer and closer to it. We say that the periodic trajectory (or the corresponding fixed point of the Poincaré mapping) *attracts* all other trajectories (points), or is *stable*.

(b)　is the reverse process: now the mapping expands everything. There's still a fixed point in the middle, but successive points move further and further away, and so do the corresponding trajectories. The periodic trajectory (or fixed point) *repels* all the others, and is *unstable*.

(c)　combines these two: it contracts in one direction but expands in another. There is again a single fixed point, which is said to be of *saddle* type: stable in one direction, unstable in the other.

(d)　is a rotation through 45°. It has a fixed point in the middle: every other point cycles through eight values, returning to where it started, because eight successive rotations through 45° combine to give the rotation through 360°, which leaves everything where it started. So every point is a period-8 point.

(e)　is only slightly more complicated geometrically, but its behaviour under iteration is much stranger. It is the *horseshoe mapping*, invented by the American topologist Stephen Smale. The original square region is stretched out long and thin, and is then

bent and replaced upon itself. Under iteration, we find an ever more complicated wiggly rectangle, which gets very long and thin and wiggles more and more. The more iterations we apply, the more complicated everything gets — the behaviour is quite different from a fixed point or a periodic point. This is an example of *chaos* — complicated and highly irregular behaviour occurring in dynamical systems. So the perfectly simple geometry of the mapping can lead to chaotic dynamics.

(*f*) the 'general' mapping can combine all of the features we have seen, and many more. Nobody knows the full range of possibilities. So a problem as simple as the iteration of a mapping in the humble plane is already beyond the reach of general mathematical understanding. The best we can do is study certain *types* of mapping and understand how each behaves.

This is a problem that is especially suited to computer explorations. To iterate a mapping by computer you just repeat the same calculation over and over again, using the end results of the previous calculation as the starting values for the next. You can monitor the progress of the calculation graphically by plotting points on the screen whose coordinates are those values. You thereby get a kind of discrete-time phase portrait.

Often, the points that you plot settle down on to some fixed object, known as an *attractor* (because they settle down to it). The attractor provides a picture of the long-term dynamics. Simple dynamics leads to simple attractors: for example, a steady state corresponds to an attractor that is a single point, a cycle of period two corresponds to an attractor made up of two points, and so on. Quasi-periodic dynamics leads to a closed loop as an attractor. But perfectly innocuous mappings, under iteration, can produce dazzlingly complex attractors, belonging to the class of geometrical objects known as *fractals*. Such dynamics is said to be *chaotic*, and it manages to be both deterministic (in principle, the future is determined forever by present conditions) and unpredictable (tiny errors increase exponentially and swamp the prediction).

To relate all this to our third theme: if the mapping has symmetry, then the attractor may be symmetric as well as fractal — a remarkable mixture of order and chaos in a single object. The relation between symmetry and chaos is just one of the topics being studied at the research frontiers.

Poincaré first encountered chaos in his work on the three-body problem. He explained why it could occur, and why that meant that there couldn't be any simple description (like Kepler's ellipses) of the motion. He didn't make much progress in understanding how to cope with chaos: that's just beginning.

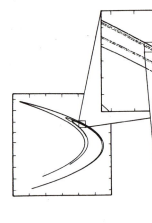

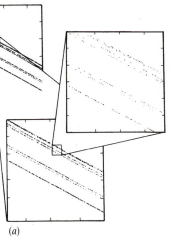

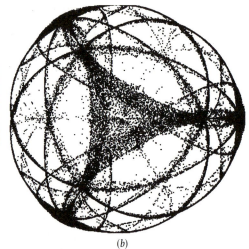

(a)   (b)

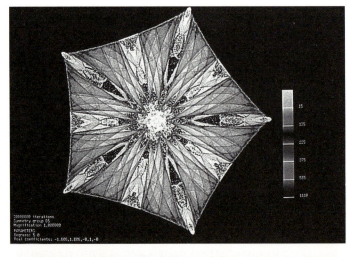

(c)

Some attractors of chaotic
attractors created by
iterating mappings:
(a) Hénon attractor;
(b) attractor with 3-fold
symmetry invented by Pascal
Chossat and Martin
Golubitsky;
(c) attractor with 5-fold
symmetry;
(d) patterned attractor defined
by a torus map, invented by
Michael Field and Martin
Golubitsky.

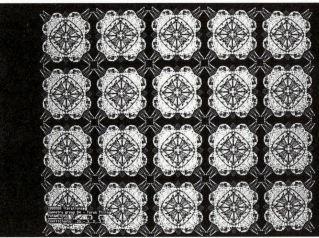

(d)

Symplectic geometry

But it seems to me that a large part [of the study of light rays] can be dealt with in a simpler and clearer fashion, if instead of calculus one makes use of some geometrical considerations, and in particular begins with a purely geometrical definition of an infinitely thin bundle of rays.

Geometrical development of the properties
of an infinitely thin bundle of rays, 1862

The equations of celestial mechanics have one special feature which makes an enormous difference to the associated dynamics: there's no air resistance in a vacuum. This means that there is no friction to slow planets down or dissipate their energy. (Strictly speaking, that's not true: there is some friction in space, caused by interstellar gas clouds, but the usual equations of motion neglect the extremely tiny frictional effect that they produce, since this only shows up on time-scales of millions of years.)

A general theory of dynamical systems in the absence of friction was developed by the Irish mathematician Sir William Rowan Hamilton, and they are generally named *Hamiltonian systems* in his honour. He discovered that the same mathematical ideas can also be applied to the behaviour of light rays in optical systems. This optical–mechanical analogy influenced many mathematicians, Möbius among them. Indeed, Möbius was one of the first to emphasize the role of geometrical methods in the study of Hamiltonian systems. Certainly Poincaré hammered the point home.

In a system without friction, energy is conserved. One (not entirely obvious) consequence of this is that a Poincaré mapping must be *volume-preserving* — that is, any tiny region of phase space retains the same (multi-dimensional analogue of) volume as it follows the dynamical flow. The imaginary 'fluid' in phase space, formed by the points that represent the states of the system, is *incompressible*.

Two-dimensional volume is just area; so when the Poincaré section has two dimensions (that is, it is a surface), the Poincaré mapping is area preserving. Here's the simplest way to construct an area-preserving mapping: form a system of concentric circles, and rotate each circle by an amount that depends continuously on the radius, keeping each circle the same size. The classical mathematicians thought that dynamics near a periodic trajectory should always look like this. Their evidence seemed quite strong: whenever they could solve the equations, that's the result they got!

All this means that near a periodic trajectory you see either

- a periodic trajectory in resonance, or

IAN STEWART

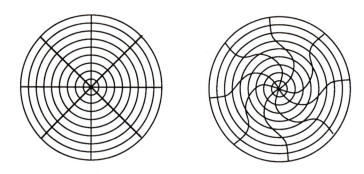

The classical picture of an area-preserving mapping involves rotating concentric circles through varying angles, so that radial lines become bent.

- a quasi-periodic trajectory — that is, something in which several separate periodic motions are combined (for example, Apollo spacecraft goes round Moon goes round Earth goes round Sun goes round Galaxy).

This follows because any nearby point sits on one of the concentric circles, and it just clicks on through a fixed angle at each iteration of the Poincaré mapping.

It's a nice, simple, attractive picture, backed up by lots of evidence. Unfortunately, it's wrong! It's a case of Catch-22. The equations can be solved only when the answer is too simple to be representative of the true picture.

Thanks to deep results of Andrei Kolmogorov, Vladimir Arnol'd, and Jurgen Moser, we now know that the typical picture near a periodic trajectory is not just a system of concentric loops. It is a far more complicated structure, known as the *vague attractor of Kolmogorov*, or VAK for short, pictured on p. 120. A point in the centre represents the original periodic trajectory. There are some closed curves surrounding it, so some quasi-periodic trajectories do exist nearby. But, sandwiched between these closed curves, there are two distinct features — smaller closed curves, forming 'island chains', and between them, a spaghetti-like tangle. At the centres of the islands are points that represent resonant periodic trajectories and the spaghetti represents chaos, so resonance and chaos are intimately linked. Each island has its own fine structure: it looks exactly like the whole VAK. So within the islands are sub-islands, mixed up with chaotic trajectories; and inside the sub-islands we find the same kind of thing; and so on, forever. It's fiendishly complicated, and nothing like the classical picture — but it's what actually happens.

The cause of this intricate structure is that harmless little requirement that the mapping should preserve area. The transformations of ordinary Euclidean geometry preserve distance: they are rigid motions. Area-preserving, or *symplectic*, transformations are much more flexible, and their natural geometry (*symplectic geometry*) is very strange to anyone brought up on a diet of Euclidean prejudices.

For instance, in symplectic geometry every line is at right angles to itself. Symplectic geometry shows up in dynamics, in quantum mechanics, and in optics: the transformation from a light source to its image in an optical system is symplectic. August Möbius was very interested in unusual geometries. He also wrote about optics, and some of his optical investigations place a similar emphasis on transformation from source to image. I think that Möbius would have approved of symplectic geometry.

The chaotic structure of the VAK, and its higher-dimensional cousins, bedevils any attempt to work on King Oscar's Prize Problem. The chaos that Poincaré encountered in his work on the three-body problem is VAK chaos. Recent calculations using a specially constructed computer, the Digital Orrery, have shown that the entire Solar System is chaotic. Among the planets, the most chaotic is Pluto: if you calculate the orbit of Pluto 200 million years ahead, you can't predict which side of the Sun it will be on. Eventually that uncertainty affects where you think the other planets will be. A billion years or so from now, Earth might be enjoying Summer when we currently predict Winter.

So although celestial chaos does occur, I wouldn't worry about it. Except perhaps for . . .

The Cosmic Cup Final

Not without some misgivings I venture to lay before the public the first fruits of my astronomical activities.

Observations from the Royal University Observatory, Leipzig, 1823

You might think that all this business of chaos and resonances and the VAK is of little consequence for us terrestrial beings. The Earth has gone round and round the Sun in orbit for four billion years or more, and life has existed for at least half that period. The long-term chaotic motion of the Solar System doesn't seem to have done us any harm . . .

The answer is: '*not yet*'. The dinosaurs, who became extinct 65 million years ago, might have had something to say about that — assuming that the theory of the great KT meteorite, whose impact with the Earth is held to have killed them off, holds up. Be that as it may, our continued existence, unperturbed (except in a gravitational sense) by the cosmos, is highly likely.

But not certain . . . At any time the universe might conspire to throw a rock at us, or a large comet or planetoid could wander in from outer space and smash us to bits. All very dramatic, the stuff of science fiction novels. Highly improbable, of course. From outer space the Earth is a pretty small target.

Yes, but not impossible: in fact, it's more likely than you might think. It isn't necessary for the universe to throw rocks at us. Our own dear Solar System is perfectly capable of doing that. The rocks are meteorites, visible in the night sky at almost any time if you watch for a while, and the culprit is Jupiter, aided and abetted by Mars. Ammunition is supplied by the asteroid belt.

Let's get a picture of the local geography. Earth is the third planet, counting outwards from the Sun; beyond are Mars (smaller and less massive than Earth) and Jupiter (huge — 11 times the Earth's diameter, and 318 times its mass). Then come four more planets. Jupiter is the largest body in the Solar System apart from the Sun, so it plays a major role in the system's dynamics. Between Mars and Jupiter are tens of thousands of smaller bodies, the asteroids (and millions of even smaller lumps of rock). Some of these manage to make their way into the Earth's orbit, enter our atmosphere, and burn up. If a really big one ever makes that trip, it will be *we* who burn up: the impact would be far worse than a hydrogen bomb.

How do asteroids enter the Earth's orbit? Why don't they stay where they are? The mechanism is intricate, and only recently understood. It begins with the possibility that an asteroid may be in 3:1 resonance with Jupiter — that is, it can go round the Sun in exactly one-third of the orbital period of Jupiter. Let's work out which asteroids do this. Jupiter orbits the Sun at a distance of 5.2 AU (astronomical units, where 1 AU is the distance of the Earth from the Sun of 150 million kilometres). Kepler's third law says that cubes of distances are proportional to the squares of the orbital periods. So an asteroid in 3:1 resonance must orbit at a distance d for which $5.2^3 : d^3 = 3^2 : 1^2$. This means that $d^3 = 5.2^3/9 = 15.623$, so that $d = 2.49$ AU. The asteroids lie in a broad belt between about 1.9 AU and 4.1 AU from the Sun, so the critical distance of 2.49 AU is well inside.

However, you don't actually *find* many asteroids around that distance. Let's see why. Such an asteroid is strongly disturbed by Jupiter, because it is repeatedly pulled in the same direction. Because of the VAK, 3:1 resonance also involves chaos. Chaos, as its name suggests, alters the asteroid's orbit in unpredictable ways. It turns out that the effect of chaos is not strong enough to send an asteroid into an Earth-crossing orbit. But it *can* send an asteroid into a Mars-crossing orbit. If Mars happens to be in the right place, the gravitational pull of Mars can fling the asteroid into an Earth-crossing orbit. That gives it a definite chance of hitting us.

It's a game of cosmic soccer: Jupiter centres the ball from a corner, Mars heads it into the terrestrial net. One fateful day the score might be Jupiter 1, Earth 0; it depends only on the random whims of chaos. It probably won't happen, of course; but it *could*. The Solar System is

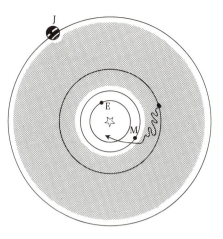

Jupiter centres, Mars scores... An asteroid in 3:1 resonance with Jupiter is subjected to repeated perturbations resulting in chaotic changes to its orbit. These can cause it to cross the orbit of Mars, and Mars can fling it inwards to encounter the Earth. Many meteorites come to Earth by this route.

a single unit, and we're part of it. Our survival for four billion years is a result of luck, rather than any preordained natural law.

The great escape

Meanwhile, for practical purposes, we have devised adequate methods, which solve the problem by approximation.

Celestial mechanics, 1843

Next, a case where all three of Möbius's major interests — symmetry, topology, and celestial mechanics — come together in a quite remarkable fashion. The question to be answered is: *just how badly behaved can the motion of a system of point particles under Newtonian gravitation be?* By 'badly behaved', I don't mean 'complicated': I'm referring to mathematical obstacles to the very *existence* of the motion.

Despite propaganda to the contrary, the differential equations of celestial mechanics do not always possess solutions that are valid for all time. Let's say that a system has a *singularity* at an instant of time if the solutions to the equations cannot be continued past that time. The simplest type of singularity in the *n*-body problem is a collision. If two bodies collide, then the equations for the system make no sense, because the force between two coincident bodies is infinite. So the equations fail to determine the motion after a collision. It's possible to wriggle out of this difficulty by insisting that the bodies bounce elastically, but even that ploy fails for a triple collision.

Are there any singularities other than collisions? For two bodies we can solve the entire problem explicitly, and the answer is 'no'. In the absence of collisions, initial conditions uniquely determine the subsequent motion for all time. In 1895, Paul Painlevé proved that the same is true for three bodies; but he couldn't extend his result to four or more bodies, and he conjectured that a non-collision singularity, or *pseudo-collision*, might occur.

What form can a pseudo-collision take? One possibility is that one or more of the bodies goes to infinity in a finite period of time! It is much easier to find such behaviour than you might think, although the easy examples do not model celestial mechanics. The simplest that I know is the differential equation

$$\frac{dx}{dt} = 1+x^2, \text{ where } x = 0 \text{ when } t = 0,$$

which has the unique solution

$$x = \tan t,$$

becoming infinite when $t = \frac{1}{2}\pi$. Thus there is a singularity at time $\frac{1}{2}\pi$ at which the value of x becomes infinite, and the solution cannot be continued past this time in any consistent or sensible fashion. Nor is there anything obviously nasty about the differential equation: $1+x^2$ is one of the simplest and best behaved functions you could think of.

The equations of motion of a system of n bodies moving under Newtonian gravity are far more complicated, but also far more special. It is not at all clear whether similar behaviour might occur, but it is also very difficult to rule it out.

Another possibility for a pseudo-collision is that one or more bodies starts to oscillate ever more wildly as time approaches some particular value. The combined researches of H. Von Zeipel, Richard McGehee, Donald Saari, and H.J. Sperling have proved that, in order to have a pseudo-collision in the n-body problem, *both* types of nasty behaviour must occur simultaneously. That is, some bodies must move off to infinity in finite time, *and* oscillate wildly. If it's going to be bad, it has to be very bad indeed. Saari also proved that pseudo-collisions in the four-body problem are infinitely rare, if they exist at all — that is, if you choose initial conditions at random, then the chance of a pseudo-collision is zero. John Mather and McGehee found a pseudo-collision for a system of four bodies confined to a line — but only after allowing an infinite number of elastic collisions first. That was interesting evidence, but not conclusive for the real problem. Then in 1984 Joseph Gerver suggested a scenario that might produce an escape to infinity, requiring the active participation of *five* bodies.

Although the bodies are point masses, I'll use astronomical imagery to remind us of their relative masses. Take three stars, one more massive than the others, and arrange them in a triangular constellation with an obtuse angle at the heaviest star. Set the star moving outwards, away from the centre of the triangle. Add a tiny asteroid that orbits round the outside of all three stars, approaching them very closely. As the asteroid passes the most massive star, arrange for it to undergo a 'slingshot' effect, gaining energy from the star, and decreasing that star's energy by the same amount. It can then, on subsequent encounters with the other two stars, transfer energy to them. As a result, the speeds of the asteroid and the other two stars can be increased. If only it were possible to speed up the heaviest star as well, the triangle could be made to expand, and the rate of expansion could be made fast enough to do the '$dx/dt = 1 + x^2$' trick and whizz off to infinity. But the law of conservation of energy prevents this.

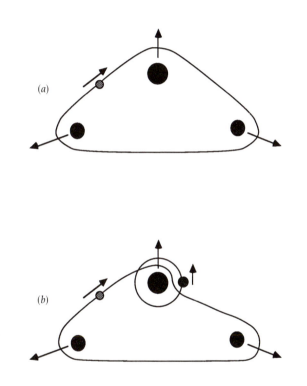

Joseph Gerver's scenario for the Great Escape:
(a) with four bodies, conservation of energy causes one star to slow down;
(b) if a fifth body is added, it can give up energy to its star, and all three stars can speed up; the extra body just orbits closer and closer to its star.

Gerver found a legal loophole. Add a *fifth* body, a planet that orbits the most massive star. Instead of making the star lose energy, make the *planet* lose energy — so much so that the star can gain some; that is, arrange the slingshot so that the asteroid whips past both planet and star, slowing the planet down but speeding up the star. Now, on each circuit of the asteroid, the stars and asteroid speed up, and the planet slows down, moving closer to its star. The energies balance, but the triangle grows — and it grows at a sufficiently fast rate that it does indeed whizz off to infinity in a finite period of time, taking the asteroid and the planet with it!

The main mathematical obstacle to proving that this works is to arrange everything so that an infinite sequence of slingshots takes place without disturbing the general scenario. The topology comes in when you try to prove that this is possible: it is used to select appropriate initial conditions. But Gerver's arguments on this point lacked rigour, and the detailed calculations became so messy that the proof couldn't be pushed through to a conclusion.

Enter symmetry: in 1989 Gerver used a clever argument, suggested by Scott Brown, to prove that the $3n$-body problem really does permit an escape to infinity for large enough values of n. Topology is heavily involved in the proof, for much the same reasons as before. Symmetry makes the calculations tractable — although still far from easy.

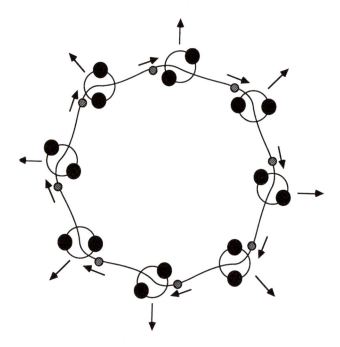

A symmetric configuration of $3n$ bodies (here $n = 8$) provides a more tractable mathematical problem. When n is sufficiently large, it can be proved that there exist initial positions and velocities for which all bodies disappear to infinity in finite time.

The configuration required is a more symmetric version of the triangle of stars. It consists of n pairs of binary stars, all having the same mass. The orbits in each pair are nearly circular, and the centres of mass of the pairs lie at the vertices of a regular n-gon. A further n planets, all having the same mass (but much smaller than the mass of the stars), move approximately along the edges of the polygon. Each time a planet approaches a binary star, it gains kinetic energy via the slingshot effect; the binary star compensates by losing kinetic energy and moving to a tighter orbit. The planet also transfers momentum to the binary star, causing it to move outwards, away from the centre of the polygon. Because of symmetry, all n

binary stars are affected in exactly the same manner at exactly the same time. At each stage,

- the polygon grows

- the planets move faster

- the binary stars close up into tighter orbits

By suitably adjusting the number of bodies, their masses, and their initial positions and velocities, it is possible to set up a system in which

- infinitely many slingshot events occur, separated by times that decrease like a geometrical progression — a, ar, ar^2, ar^3, ..., with $r < 1$,

- the polygon grows by at least a fixed amount k at each slingshot,

- the shape of the configuration of $3n$ bodies retains the same basic characteristics.

The rigorous proof uses topology to show that the entire infinite sequence of events can be made to take place. Then infinitely many slingshots occur during a total time of

$$a + ar + ar^2 + ... = a/(1 - r),$$

which is finite. Meanwhile the polygon has grown by a size ∞k, which is infinite. So the entire system escapes to infinity in finite time. Moreover, the encircling binary stars oscillate more and more wildly as their orbits shrink to zero radius.

The role of symmetry is to simplify the calculations enough to make a proof possible. Effectively, it reduces the problem from $3n$ bodies to three. Once we have determined the positions and velocities of one binary star and one planet, we apply the n-fold rotational symmetry to find those of the remaining $3n - 3$ bodies. Thus we have only to keep track of three bodies, and the rest is taken care of by symmetry. In other words, we can reduce the problem to a different problem about three 'bodies', each an n-gon of point masses, and moving under rather complicated forces. This problem is (just) tractable, at least for enough n when the forces simplify, and Gerver managed to complete the proof.

I should mention that in 1988, Z. Xia proved that the five-body problem has a solution in which all five bodies escape to infinity in finite time. His scenario is different from Gerver's, but it also involves symmetry, and a topological proof. I must also emphasize that this is a purely mathematical result. In the real world, particles are never points, so the binary stars would eventually touch and the scenario would break down. Moreover, the real universe is relativistic, with velocities limited by the speed of light, so nothing can *ever* escape to infinity in finite time. (Is *that* why the deity made the universe relativistic?) Indeed, the universe is probably finite anyway. It's a mathematical fact about a particular system of differential equations that *model* the real world. But a remarkably curious one . . .

Broken symmetry

So, by motion of one figure, the totality of its points can be brought into coincidence with the corresponding points of the other.

On symmetrical figures, 1851

The importance of symmetry in dynamics is becoming ever more apparent. In particular, a mechanism known as *symmetry-breaking* is known to be responsible for many of the patterns found in nature. Symmetry-breaking occurs when a symmetrical system behaves in a less symmetrical way.

You carry one example around with you all the time — or rather, *it* carries *you*. Human beings are (approximately) left–right symmetric. If this symmetry always held, then when you walked, your left and right legs would both move forward at the same time: *BUMP*! So walking breaks the bilateral symmetry of the human body. However, some symmetry remains: each leg does the same thing except for a time delay — your left leg is half a step behind (or in front of) what your right leg is doing. Many animal gaits (trot, gallop, bound) have symmetries that break the symmetry of the stationary animal, and they all involve this kind of time delay.

Symmetry-breaking is common on Earth, and it provides a unified framework for the study of pattern formation. There are patterns in the heavens too, and symmetry-breaking on the grandest of scales is involved in one of the most beautiful celestial patterns of all.

Across the night sky there runs a huge, irregular luminous band — the Milky Way. Though mysterious to the naked eye, the Milky Way reveals its true nature to any small telescope: it is composed of huge numbers of faint stars. Around the end of the eighteenth century, the astronomer William Herschel demonstrated that the apparent density of stars is greatest along the Milky Way, and decreases steadily away from it. In the mid-nineteenth century, the German philosopher Immanuel Kant and the Englishman Thomas Wright deduced that the system of stars forms a flattened disc, with the Sun inside. The argument is simple. If our own Sun, and with it the Earth from which we observe the heavens, are buried within a disc of stars, then more stars will be seen in directions parallel to the disc than at right angles to it. This hypothetical disc of stars was humanity's first inkling of the existence of the structure now known as a *galaxy*. It soon became clear that the universe is composed of innumerable galaxies, and that each galaxy is composed of innumerable stars.

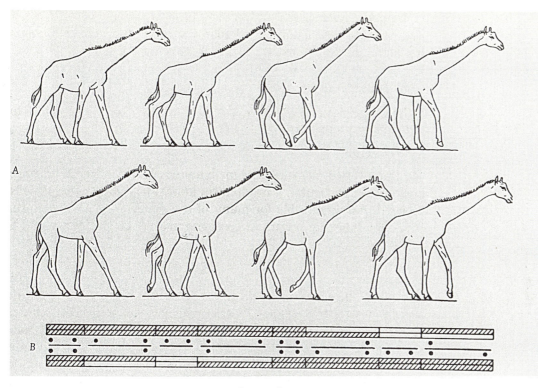

A

B

The walk of the giraffe is typical of the kind of symmetry that occurs in animal gaits, involving both spatial and temporal transformations. The second row of pictures, occurring half a period later than the first row, is its mirror image.

Many galaxies have internal structure, and the most dramatic are magnificent whirlpools of light — multi-armed spirals like a Catherine wheel. The American astronomer Edwin Hubble classified galaxies into four main types:

- *elliptical*, smooth featureless blobs containing little gas or dust, generally lacking a sharp outer edge

- *lenticular*, prominent disc, containing no gas, no bright young stars and no spirals — distribution of brightness similiar to that of spirals

- *spiral*, prominent disc, including gas and dust as well as stars — spiral arms, of somewhat variable geometry, some tightly wound, others long and thin

- *irregular*, the rest

These four types of galaxies merge continuously into a single sequence, the *Hubble sequence*:

$$elliptical \rightarrow lenticular \rightarrow spiral \rightarrow irregular.$$

The most dramatic structure, symmetry on an enormous scale, occurs in spiral galaxies. Perfect m-armed spirals are symmetric under rotation through an angle of $360°/m$. Galaxies are not perfect spirals, but m-fold rotational symmetry is often present to a good

Broken symmetry on a cosmic scale: the dramatic spirals of M51, the Whirlpool Galaxy.

degree of approximation. For spiral galaxies, by far the commonest case is $m = 2$, and two-armed spirals are unchanged if you rotate them through 180°.

Why do galaxies exhibit spirals? Until the early 1960s, virtually all astronomers thought (wrongly) that the spiral arms were caused by the inter-stellar magnetic field. Only one, the Swede Bertil Lindblad, had the right idea: that spirals are purely dynamic in origin, caused by gravitational interaction. In 1965, C. C. Lin and Frank Shu had the crucial insight that crystallized Lindblad's ideas into a well-formed theory. Until then, there had been a tendency to assume that the spiral arms are fixed features, in the sense that a particular star is either in a spiral arm, or not, and as time passes, it stays there. As

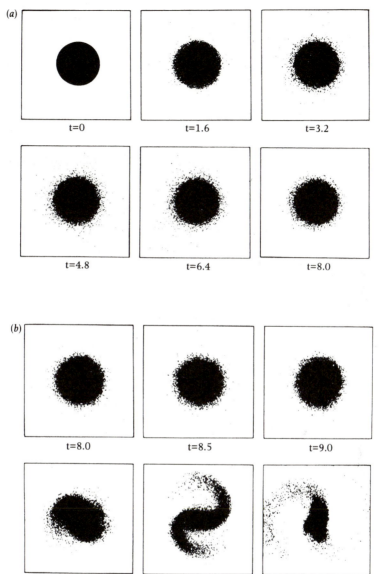

Numerical simulations of galaxy dynamics due to F. Hohl. The initial state in both cases is a circular disc of stars:
(a) if circular symmetry is imposed then the disc just becomes fuzzy at the edges;
(b) if the symmetry is allowed to break, spirals appear.

the galaxy revolves, the arms revolve with it, and so do the stars in the arms. This belief poses enormous problems if you think that only gravitational interactions are relevant, because a rigidly rotating galaxy should tear itself apart by centrifugal force.

Lin and Shu realized that there is another possibility: the spiral structures might be *density waves*, sweeping through the system of stars. The wave can maintain its overall coherence and structure, even if the stars move into and out of the regions of greatest density — the arms. This is easy to understand: we see examples every day. Take a length of rope, tie it to a wall, and move the other end up and down, sharply. A wave travels along rope from your arm, bounces off the wall and returns. But do *individual bits of rope* travel along to the wall and back? Of course not!

Density waves alone aren't enough to solve the problem of spiral structure, so Lin and Shu also assumed that the spiral density wave is a kind of steady state. If you ignore the rotation of the galaxy, the pattern of spirals will always look pretty much the same. It's such a simple theory that it looks almost obvious. With a large dose of 20–20 hindsight, we can interpret the Lin–Shu theory as an example of symmetry-breaking. The crucial point is that a rotating spiral structure has a lot of symmetry — but less than a rotating disc has.

Indeed, symmetry-breaking is apparent in some of the earliest numerical experiments on galaxy dynamics. In 1971, F. Hohl used a computer to simulate the motion of a disc containing 100 000 stars. To begin with, he assumed that the stars are uniformly distributed inside a circular region, and he simulated the motion on the assumption that *the circular symmetry remains unbroken*. The result is straightforward: the disc becomes fuzzy at the edges, but remains circular in form. When the assumption of unbroken circular symmetry is relaxed, however, the results are different and striking. At first, the disc becomes fuzzy; but then it starts to bulge on opposite sides and a pair of spiral arms develops, linked by a central bar. Thus the rotating disc undergoes a symmetry-breaking instability. Not all symmetry is lost, however: the two-armed structure has a single rotational symmetry, through 180°.

Let's interpret these results from the point of view of symmetry-breaking. Begin with a rotating circular disc of stars, in equilibrium. This is a system with circular symmetry — unchanged by rotation through all angles. If the symmetry breaks, then the general theory of symmetry-breaking tells us that the only possibilities are m-fold symmetry — rotation through angles that are multiples of $360°/m$ — for some integer value of m. Thus we can 'postdict' (predict after the fact) the observed symmetries of spiral galaxies.

The detailed spiral form is beyond the reach of pure symmetry argument, and its calculation depends upon physical models such as Hohl's; but m-armed spirals are the *simplest* structures with m-fold rotational symmetry, so it's hardly a surprise if they're what you see. The predominance of two-fold symmetry also depends upon the detailed model, but a rough rule of thumb, based on experience in many different special cases, is that m-fold instability is common for the smaller values of m, and $m = 2$ is the smallest that can occur here.

In other words, very general considerations of dynamics and symmetry — simple, fundamental mathematical principles — go a long way towards explaining what was previously a deep mystery.

The spiral arms of galaxies are the visible relics of broken dynamical symmetry on a cosmic scale.

It would have pleased Möbius no end.

The great plodder

It is my intention, through the methods and theorems of geometry herein laid down, to contribute in some measure to the simplification of its investigations and the broadening of its horizons...

The barycentric calculus, 1827

Some mathematicians are remembered for deep and far-reaching discoveries: Isaac Newton and Gottfried Leibniz for the calculus; Joseph Fourier for his application of trigonometrical series to the theory of heat; Nikolai Ivanovitch Lobachevsky and János Bolyai for non-Euclidean geometry; Ferdinand Lindemann for the transcendence of π; Carl Friedrich Gauss for the law of quadratic reciprocity. Others have great theorems, or great theories, named after them: Hilbert's basis theorem, Pythagoras's theorem, the Schröder–Bernstein theorem, Zorn's lemma, Galois theory.

There's a Möbius band, function, transformation, net, and inversion formula. Not bad — but there's no Möbius's theorem. There *could* have been a Möbius theory, mind you: it would have been a perfectly reasonable name for his barycentric calculus. In fact, if he'd given that theory a more cumbersome or less apt designation, mathematicians might have been forced to name it after him in order to remember what it was.

Historical recognition is at best a fickle thing, but in this case its absence is appropriate. The reason lies in Möbius's mathematical style. He was a top-rank mathematician, but not a spectacular one. He spent his academic career as a professor of astronomy — and the work that he did in that subject is competent, but routine. Even in his true calling, mathematics, he was not so much an original creator as a formalizer and simplifier. To put it bluntly, he was a bit of a plodder; but when Möbius plodded he plodded with diligence, elegance, and imagination. He never stopped, and he *got* places. His great talent was sorting out other people's ideas and seeing them clearly — often more clearly than their creators had done. He appreciated the need to place mathematical ideas in a natural formal setting. He would have understood immediately the modern mathematician's cry, so baffling to those scientists who think that solving a problem is the end of it: 'what's *really* going on here?'

It is an accident of history that his name is remembered because of a topological party-piece. But it was typical that Möbius should notice a simple fact that anyone could have seen in the previous two thousand years — and typical that nobody did, apart from the simultaneous and independent discovery by Listing. But Möbius's importance for mathematics does not rest upon such shallow foundations. What makes a great mathematician? A feel for form, a strong sense of *what is important*. Möbius had both in abundance. He *knew* that topology was important. He *knew* that symmetry is a fundamental and powerful mathematical principle. The judgement of posterity is clear: *Möbius was right*. His mathematical taste was imaginative and impeccable. And, while he may have lacked the inspiration of genius, whatever he did he did well, and he seldom entered a field without leaving his mark.

No body of deep theorems . . . but a style of thinking, a working philosophy for doing mathematics effectively and concentrating on what's important. That is Möbius's modern legacy. We couldn't ask for more.

Notes on contributors

Norman Biggs is Professor of Mathematics at the London School of Economics. His major interest is in discrete mathematics, and he has published several books and many research papers in this area. For many years he has also worked on the history of mathematics, and among his publications in this field is *Graph theory 1736–1936* (1976), written in conjunction with Keith Lloyd and Robin Wilson.

Allan Chapman teaches the history of science at the University of Oxford, where he is attached to Wadham College. His interests lie in the history of astronomy, with particular emphasis on the development of astronomical instruments and observatories. He has edited J. Flamsteed's *Historia* (1982), and written *Dividing the circle* (1990).

John Fauvel is Senior Lecturer in Mathematics at the Open University, and President of the British Society for the History of Mathematics. He has been an editor of several books, including *Darwin to Einstein: Historical studies on science and belief* (1980), *Conceptions of inquiry* (1981), *The history of mathematics: a reader* (1987), and *Let Newton be!* (1988).

Raymond Flood is University Lecturer in Computing and Mathematics at the Department for Continuing Education, Oxford University, and Vice President of Rewley House. His main research interests lie in mathematical statistics. He is co-editor of *The nature of time* (1986) and *Let Newton be!* (1988).

Jeremy Gray is Senior Lecturer in Mathematics at the Open University. His main research area is the history of mathematics in the nineteenth century, particularly the growth of complex function theory and algebraic geometry. He is the author of several books, including *Ideas of space: Euclidean, non-Euclidean, and relativistic* (1989), and co-edited (with John Fauvel) *The history of mathematics: a reader* (1987).

Gert Schubring is a member of the Research Institute for Didactics of Mathematics at the University of Bielefeld. His interests focus on nineteenth-century science and mathematics, especially on comparative developments in France and Germany. He has published on the interaction between conceptual and institutional factors in the development of mathematics and the sciences.

Ian Stewart is Professor of Mathematics at the University of Warwick. His research area is bifurcation theory and non-linear dynamics. He is active in the popularization of mathematics, and is the author of over fifty books, including *The problems of mathematics* (1987) and *Does God play dice?* (1989). He writes regularly on mathematical issues, and contributes to the Mathematical Games column of *Scientific American*.

Robin Wilson is Senior Lecturer in Mathematics at the Open University. He has written and edited a number of books on graph theory and combinatorics, including *Introduction to graph theory* (1972) and *Selected topics in graph theory* (1978). He is increasingly involved with the popularization of mathematics and with the history of mathematics, and is a co-editor of *Let Newton be!* (1988).

References

This list gives detailed references to the quotations in the book. For convenience, we use the abbreviation *Werke* for the following frequently cited work:

August Möbius, *Gesammelte Werke*, 4 volumes (edited by R. Baltzer, F. Klein, and W. Scheibner), Leipzig, 1885–7; reprinted, Dr. Martin Sändig oHG, Wiesbaden, 1967.

CHAPTER 1

7   'I find it utterly impossible'   Heinrich von Treitschke, *History of Germany in the nineteenth century: selections*, Chicago University Press, 1975, p. 151.

10   'The inspirations for his research'   *Werke* Vol. 1, pp. v–xx.

14   'There was once a king'   N.L. Biggs, E.K. Lloyd, and R.J. Wilson, *Graph theory 1736–1936*, Clarendon Press, Oxford, 1976, pp. 115–16.

CHAPTER 2

21   Table of names *Verzeichniss jetzt lebender Mathematiker von Rang*, Nachlass G.S. Ohm, Deutsches Museum, Abteilung Sondersammlungen, No. 67.

29   'There I had the courage'   W. Ahrens, *Briefwechsel zwischen C.G.J. Jacobi und M.H. Jacobi*, Teubner, Leipzig, 1907, p. 90.

29   'We are unhappy'   L.K. Königsberger, *C.G.J. Jacobi. Festschrift zur Feier der 100. Wiederkehr seines Geburtstages*, Teubner, Leipzig, 1904, pp. 133–4.

30   'In this way'   ibid., p. 134.

31   'So it is also important'   G. Schubring, Pläne für ein Polytechnisches Institut in Berlin, in F. Rapp and H.W. Schütt (eds.) *Philosophie und Wissenschaft in Preussen*, Technische Hochschule, Berlin, 1982, p.216.

CHAPTER 3

46   'it was, therefore, thrown out'   John F.W. Herschel, *Astronomy*, London, 1833, p. 277.

58   '... an idiot, with a few days' practice'   G.B. Airy, in an official letter of 1847. Airy Archives, RG 06, 2/293. Royal Observatory Archives, Cambridge University Library.

65   '... surrounded by a gaseous atmosphere'   H.C. King, *History of the telescope*, Charles Griffin & Co., London, 1955, p. 283.

CHAPTER 6

121   'While holding the side *AB*'   *Werke*, Vol. II, p. 520.

122   'Two geometrical figures'   *Werke*, Vol. II, p. 435.

122   The figures are taken from   *Werke*, Vol. II, p. 520.

123   'A figure is symmetric'   *Werke*, Vol. II, p. 353.

124   The figure is taken from   *Werke*, Vol. II, p. 688.

125   'By the law of universal'   *Werke*, Vol. IV, p. 107.

125   A potted history of the development of celestial mechanics is given in Chapter 13 of Ian Stewart, *The problems of mathematics*, Oxford University Press, 1987.

125   'Kepler's First Law'   *Werke*, Vol. IV, p. 78.

126   For readings from Kepler, and more about his polyhedral theory of planetary distances, see D.L. Hurd and J.J. Kipling (eds.), *The origins and growth of physical science*, Penguin, Harmondsworth, 1964.

127   'A complete solution'   *Werke*, Vol. IV, p. 155.

128   For a discussion of Poincaré's ideas about celestial mechanics, see Chapter 4 of Ian Stewart, *Does God play dice?*, Blackwell, Oxford, 1989; Penguin, Harmondsworth, 1990.

129   'Each new system of axes'   *Werke*, Vol. I, p. 7.

132   'The circle itself'   *Werke*, Vol. II, p. 332.

133   'A method will be developed'   *Werke*, Vol. II, p. 440.

136   'A real-valued function'   *Werke*, Vol. II, p. 445.

139   'It is not difficult'   *Werke*, Vol. IV, p. 175.

141   'But this is not possible' *Werke*, Vol. II, p. 82.
        For further information about chaos see James Gleick, *Chaos: making a new science*, Viking Press, New York, 1987, and Ian Stewart, *Does God play dice?*, Basil Blackwell, Oxford, 1989; Penguin, Harmondsworth, 1990.

145   'But it seems to me that'   *Werke*, Vol. IV, p. 571.

146   The problem of the stability of the Solar System is described in J. Moser, 'Is the Solar System stable?', *Mathematical Intelligencer*, **1** (1978), 65–71. Chaos in the Solar System is the subject of Chapter 12 of Ian Stewart, *Does God play dice?*, Basil Blackwell, Oxford, 1989; Penguin, Harmondsworth, 1990. Technical details may be found in the article by Jack Wisdom in *Dynamical chaos* (edited by M.V. Berry, I.O. Percival, and N. Weiss), Royal Society, London, 1988.

147   'Not without some misgivings'   *Werke*, Vol. IV, p. 457.

149   'Meanwhile, for practical purposes'   *Werke*, Vol. IV, p. 155.
        The great escape is discussed in J.L. Gerver, 'A possible model for a singularity without collisions in the five body problem', *Journal of Differential Equations*, **52** (1984), 76–90; and J.L. Gerver, *The existence of pseudocollisions in the plane*, Rutgers University, 1989.

154   'So, by motion of one figure'   *Werke*, Vol. II, p. 569.
        For animal gaits and symmetry, see P. Gambaryan, *How mammals run: anatomical adaptations*, Wiley, New York, 1974; and J.J. Collins and Ian Stewart, *Coupled nonlinear oscillators and the symmetries of animal gaits*, University of Warwick, 1990.

155   Many pictures of galaxies are given in Paul Murdin and David Allen, *Catalogue of the Universe*, Cambridge University Press, 1979.

157   The dynamics of galaxies, and in particular the formation of spiral arms, is described in James Binney and Scott Tremaine, *Galactic dynamics*, Princeton University Press, 1987.

159   'It is my intention'   *Werke*, Vol. I, p. 12.

Picture credits

Frontispiece: Möbius band II, woodcut, printed from three blocks, 1963, Courtesy of Cordon Art-Baarn-Holland.

Facing p. 1: Frontispiece of Möbius *Werke*.

2   Archiv Pförtner-Bund, Meinerzhagen.

3   From Franz Mehring, *Absolutism and Revolution in Germany 1525–1848*, New Park Publications, 1975.

4   Engraving by Johann Georg Schreiber, Courtesy of John Fauvel.

5   From T. Wallbank, A. Taylor, G. Carson Jnr, M. Mancall, *Civilization Past and Present*, Scott Foresman, 1969.

6   Contemporary painting by C. G. H. Geissler, courtesy of John Fauvel.

8   *Festschrift zur 500-Jähr-Feier Leipziger Universitäts*, Vol. IV. Tafel II. 1909, Bodleian Library Oxford.

8   Möbius, *Werke*, IV, p. 446.

11   Möbius, *Werke*, IV, title page.

12   Möbius, *Werke*, IV, p. 594.

14   Möbius, *Werke*, IV, p. 638.

15   Cover illustration by Steve Chalker from Martin Gardner, *The No-Sided Professor*, Prometheus Books, Buffalo, New York, 1987.

16   Cover illustration from Francis Schiller, *A Möbius Strip*, University of California Press, 1982.

17   Frontispiece from Francis Schiller, *A Möbius Strip*, University of California Press, 1982.

17   Paul Möbius, *Ausgewähte Werke*, 1905, Vol. 7, Tafel III, The British Library.

20   *Annals of Science*, 33, 1976.

22   Hans-Heinrich Himme, *Stich-haltige Beiträge zur Geschichte der Georgia Augusta in Göttingen*, Vandenhoeck und Ruprecht, 1987.

23   Hans Hübner, *Geschichte der Martin-Luther-Universität Halle-Wittenberg 1502–1977*, 1977.

23   Hans-Heinrich Himme, *Stich-haltige Beiträge zur Geschichte der Georgia Augusta in Göttingen*, Vandenhoeck und Ruprecht, 1987.

24   Courtesy of Johannes Karsten.

25   Courtesy of Gert Schubring.

26   Courtesy of Gert Schubring.

27   *Festschrift zur 500-Jähr-Feier Leipziger Universitäts*, Vol. IV. Tafel II. 1909, Bodleian Library Oxford.

28   Frontispiece of C. G. J. Jacobi, *Gesammelte Werke*, edited by C. W. Borchardt, Berlin, 1881.

29   Frontispiece of G. P. L. Dirichlet, *Werke*, edited by L. Kronecker, Berlin, 1889.

30   Title page of first issue of *Journal für die reine und angewandte Mathematik*, 1826.

34   Courtesy of the Museum of the History of Science, Oxford.

36   From D. Botting, *Humboldt and the Cosmos*, Sphere Books, 1973.

38   Courtesy of the Museum of the History of Science, Oxford.

39   Hans Kraemer, *Das XIX Jahrhundert in Wort und Bild*, Berlin, 1900.

40   Möbius, *Werke*, IV, p. 639.

41   From J. Ashbrook, *Astronomical Scrapbook*, Cambridge University Press, 1988.

41   *Publikationen Kaiserlichen Universitäts-Sternwarte*, 24, Part I, 1914.

42   Courtesy of the Museum of the History of Science, Oxford.

44   W. Pearson, *Astronomy*, 1829. Courtesy of the Museum of the History of Science, Oxford.

47   W. Pearson, *Astronomy*, 1829. Courtesy of the Museum of the History of Science, Oxford.

48   A. Von Schweiger-Lerchenfeld, *Atlas der Himmelskünde*, 1898, Radcliffe Science Library Oxford.

51   From J. Ashbrook, *Astronomical Scrapbook*, Cambridge University Press, 1988.

52   Portrait by J. Russell, 1794, courtesy of the Science Museum London.

53   Dorritt Hoffleit, *Some Firsts in Astronomical Photography*, Cambridge, Mass, 1950.

54   Frontispiece of F. W. Bessel's collected works, edited by R. Engelmann, Leipzig, 1875. Courtesy of the Museum of the History of Science, Oxford.

55   Dorritt Hoffleit, *Some Firsts in Astronomical Photography*, Cambridge, Mass, 1950.

59   *L'Illustration*, Paris, 8, 1846.

60   Courtesy of the Royal Society.

60   John Herschel's photography courtesy of the Science Museum London.

61   *Philosophical Transactions* IXXIV, 1784 and IXXV, 1785.

62   From R. S. Ball, *The Starry Heavens*, London, 1905.

63   From T. E. R. Phillips, *Splendour of the Heavens*, Hutchinson, 1923.

63   From J. N. Lockyer, *Stargazing*, London, 1878.

64   From T. E. R. Phillips, *Splendour of the Heavens*, Hutchinson, 1923.

64   Spectroscopic apparatus from J. N. Lockyer, *Stargazing*, London, 1878.

65   From J. N. Lockyer, *Stargazing*, London, 1878.

66   From J. Ashbrook, *Astronomical Scrapbook*, Cambridge University Press, 1988.

67   Johann Schröter, *Aerographische Beiträge*, Leiden, 1881.

69   J. A. Repsold, *Astronomische Meßwerkzeurge*, Munich, 1908. Courtesy of the Museum of the History of Science, Oxford.

70   J. A. Repsold, *Astronomische Meßwerkzeurge*, Munich, 1908. Courtesy of the Museum of the History of Science, Oxford.

72   A. Von Schweiger-Lerchenfeld, *Atlas der Himmelskünde*, 1898, Radcliffe Science Library Oxford.

74   From J. N. Lockyer, *Stargazing*, London, 1878.

78   Möbius, *Werke*, I, title page.

80   Courtesy of the Ecole Polytechnique, Paris.

92   Frontispiece of Julius Plücker, *Gesammelte wissenschaftlichen Abhandlungen*, Leipzig, 1895-6.

93   Möbius, *Werke*, III, title page.

102   Möbius, *Werke*, I, p. 172.

104 Stamp courtesy of Robin J. Wilson.

106 Stamp courtesy of Robin J. Wilson.

109 From N. L. Biggs, E. K. Lloyd, R. J. Wilson, *Graph Theory 1736–1936*, Clarendon Press, Oxford, 1986.

110 J. B. Listing, *Vorstudien zur Topologie*, 1847.

111 J. B. Listing, *Der Census räumlicher Complexe*, 1861.

112 Möbius, *Werke*, II, p. 484.

113 Stamp courtesy of Robin J. Wilson.

114 From N. L. Biggs, E. K. Lloyd, R. J. Wilson, *Graph Theory 1736–1936*, Clarendon Press, Oxford, 1986.

118 Stamp courtesy of Robin J. Wilson.

120 R. Abraham, J. E. Marsden, *Foundations of Mechanics*, Addison-Wesley Publishing Company Inc., 1978.

122 Möbius, *Werke*, II, p. 520.

124 Möbius, *Werke*, II, p. 688.

125 Möbius, *Werke*, IV, p. 68.

126 I. Stewart, *Does God Play Dice?*, Basil Blackwell, 1989.

128 I. Stewart, *Does God Play Dice?*, Basil Blackwell, 1989.

129 Courtesy of Ian Stewart.

130 I. Stewart, *Does God Play Dice?*, Basil Blackwell, 1989.

133 Courtesy of Ian Stewart.

134 Courtesy of Ian Stewart.

135 Courtesy of Ian Stewart.

136 Möbius, *Werke*, II, p. 445.

137 Courtesy of Ian Stewart.

139 Courtesy of Ian Stewart.

142 Courtesy of Ian Stewart.

144 (a), (c) courtesy of Ian Stewart.

144 (b), (d) M. Field, M. Golubitsky, *Symmetry in Chaos*, Oxford University Press, 1992.

149 Courtesy of Ian Stewart.

151 Courtesy of Ian Stewart.

152 Courtesy of Ian Stewart.

155 P. P. Gambaryan, *How Mammals Run*, Kuperard, London.

156 P. Murdin, D. Allen, *Catalogue of the Universe*, Cambridge University Press, 1979.

157 J. Binney, S. Tremaine, *Galactic Dynamics*, Princeton University Press, 1987.

Index of names